Titles in This Series

Titles in This Series

Titles in This Series

Partition Problems in Topology

CONTEMPORARY
MATHEMATICS

Volume 84

Partition Problems in Topology

Stevo Todorcevic

AMERICAN MATHEMATICAL SOCIETY
Providence · Rhode Island

EDITORIAL BOARD

1980 *Mathematics Subject Classification* (1985 *Revision*). Primary 04-02, 03E05, 03E50; Secondary 50-02, 54A25.

Library of Congress Cataloging-in-Publication Data

Todorcevic, Stevo.
    Partition problems in topology/Stevo Todorcevic.
    p. cm. –(Contemporary mathematics, ISSN 0271-4132; v. 84)
    Bibliography: p.
    Includes indexes.
    ISBN 0-8218-5091-1
    1. Topology. 2. Set theory. I. Title. II. Series.
QA611.T63   1989
514–dc 19

88-39032
CIP

CB.SLT = Math
SCIMON S4

CONTENTS

PREFACE

This book deals with a topological version of the Ramsey problem for the uncountable. Simple cases of this problem include, for example, the famous Souslin problem and the well-known problem from topological measure theory asking if all regular Radon measures are σ-finite. The essence of the problem is combinatorial and set-theoretical, but unlike most of the problems of this type (such as the Whitehead problem from group theory or the problem of automatic continuity from the theory of Banach algebras) the set-theoretical methods needed for the solution of the problem have not yet been developed. One of the reasons for this is that the basic notions of set theory such as the notion of cardinality or the notion of a stationary set seem to be quite irrelevant to the problem.

The book is the result of rewriting a set of handwritten notes which were produced during the last few years and distributed informally to a large number of experts in the area. My special thanks are due to Liz Stimmel of the University of Colorado for the excellent typing.

Boulder,
May 1988

INTRODUCTION

In this memoir we present results closely related to the well-known conjecture of general topology asserting that the following two properties are equivalent for every regular space X:

(HS) Every subset of X contains a countable subset with the same closure.

(HL) Every family of open subsets of X contains a countable subfamily with the same union.

The statement that (HS) implies (HL) will be denoted by (S) and the statement that (HL) implies (HS) will be denoted by (L). In the vast literature on this subject these two statements are usually called the "S-space problem" and the "L-space problem," respectively. The separate consideration of the two implications may seem unnatural at first, but one of the results of this memoir is that (S) and (L) are actually quite different problems and that the methods designed for solving one of them do not work for the other.

The statements (S) and (L) are claiming the equivalence of two quite diverse topological properties and *a priori* one sees no reason why they should be true. But the problems seem to be very basic for general topology because they are connected to many subjects of this field (see [21], [37], [41]). For example, one can hardly find an expert in this field not having a result touching (S) and (L). Quite often when working on seemingly unrelated problems in their field they encounter a difficulty which can be resolved if one knows the answer to certain instances of (S) and (L). Even more often, the proof technique designed for solving (S) and (L) turns out to be useful in many other problems of general topology. Why is this so? Well, this is so because Ramsey-type theorems are basic and so much needed in many parts of mathematics and (S) and (L) happen to be Ramsey-type properties of the uncountable most often needed by the general topologist. To understand this point one only needs to remember that (S) and (L) have simple translations (see §7) asserting the existence of uncountable homogeneous sets for certain partitions of $[\omega_1]^2$ into two sets. This also explains the reason why these two topological questions demand so much from set theory for their solution. In

1

fact, (S) and (L) are the deepest set-theoretical questions we know of which originated in a field outside of set theory.

In §§0-3 we show that the countability requirements in the statements of (S) and (L) are essential. In particular, we shall present regular spaces having different hereditary separability and hereditary Lindelöfness. This question has been quite often asked in the literature (see e.g., [10; pp. 155, 284], [41; Question (5)]). In §4 we present (or mention) small ccc nonseparable compact spaces and also examples of $sup \neq max$ for of certain cardinal functions such as the spread. The former example are relevant to (S) and (L) because they give answers to certain strong forms of Souslin Hypothesis which is easily seen to be a special form of the Conjecture (L). The examples concerning the negative answer to the $sup = max$ problem for the cardinal function spread quite often contain subspaces relevant to (S) and (L).

In §5 we explain the relationship between (S) and (L) and the Souslin Hypothesis which is a consequence of both. The §6 is about Luzin sets and Luzin spaces and their connections with (S) and (L). This section is telling us that CH-diagonalization arguments, so much used in the previous work on the subject, have produced examples which can also be obtained from the most simple and early CH-diagonalization arguments – those producing Luzin sets of reals.

In §§7 and 8 we present certain forcing axioms in forms of Ramsey-type properties of the uncountable in order to study (S), (L) or questions closely related to them. It is this subject which will require further study if we want to answer even the consistency questions of (S) and (L).

In §9 we prove that (S) and (L) are different statements contrary to expectations of many experts.

Most sections require minimal knowledge of set theory and topology. They contain some simple-looking arguments whose proper understanding, however, might require some closer study. The sections 8 and 9 will assume a familiarity with the papers [54], [55], [57] and [0].

Most of the results of this monograph have been circulated as handwritten notes over the last few years. They are listed in [62] in the chronological order. Some historical remarks will be added at the end of each section.

All spaces in this monograph are assumed to be Hausdorff. The notation will be fairly standard. The only possible exceptions might be the following. If f is a function and A a set, then by $f \mid A$ we denote the restriction of f on A. By $f''A$ we denote the set of all $f(a)$ for a in A. The function f will frequently be identified with its graph, i.e., the set of all ordered pairs $\langle x, y \rangle$ such that $f(x) = y$. By dom(f) we denote the domain of f and by range (f) its range, i.e., the set $f''A$ for $A = \text{dom}(f)$.

0. THE ROLE OF COUNTABILITY
IN (S) AND (L)

In the first four sections of this monograph we shall show that "the unusual, and seemingly unnecessary, position of denumerability in topology" ([65], [64; p. 83]) appears to be quite essential in the statements of (S) and (L). We shall present there several examples of spaces showing that some high-level analogues of (S) and (L) are false. The examples will also show a strong influence of (S) and (L) on the set of real numbers.

Let X be a space and let $\leq$ be a reflexive and transitive relation on X. Then by $X[\leq]$ we denote the space obtained by refining the topology on X by introducing the sets

$$\{y \in X : y \leq x\}, \ (x \in X)$$

as new open sets. The following is a list of few straightforward facts about $X[\leq]$ which give us some indications which properties of X and $\leq$ are needed in order to make $X[\leq]$ relevant to (S) and (L).

0.0. LEMMA. *The character of the space $X[\leq]$ is not bigger than the character as X.*

PROOF. If B is an open neighborhood of x in X, then

$$B[\leq x] = \{y \in B : y \leq x\}$$

is an open neighborhood of x in $X[\leq]$. Moreover, if $\mathcal{B}_x$ is a basis of x in X, then

$$\{B[\leq x] : B \in \mathcal{B}_x\}$$

is a basis of x in $X[\leq]$.

0.1. LEMMA. *If X is regular (zero-dimensional) and if $\leq$ is closed in X^2, then $X[\leq]$ is also regular (zero-dimensional).*

PROOF. If $\leq$ is closed in X^2, then for all open $B \subseteq X$ and x in X, $B[\leq x]$ is the intersection of an open set and a clopen set. Note also that the closure of $B[\leq x]$ in $X[\leq]$ is a subset of $\bar{B}[\leq x]$. Now the conclusion is immediate.

3

0.2. LEMMA. *If D is a discrete subspace of $X[\leq]$, then D is the union of $w(X)$ antichains of $X, \leq$.*

PROOF. If B is a neighborhood of both x and y, then

$$x \notin B[\leq y] \text{ and } y \notin B[\leq x]$$

imply that x and y are incomparable. Using this fact the conclusion of 0.2 follows immediately.

Recall that a space Y is *right (left)-separated* iff there is a well-ordering $<_w$ on Y and for each y in Y a neighborhood U_y such that $x \notin U_y$ for all $x >_w y$ ($x <_w y$). The reason for introducing these notions is that for any space Z, the hereditary density of Z, $hd(Z)$, is equal to the supremum of cardinalities of left-separated subspaces of Z, and that the hereditary Lindelöf number of Z, $hL(Z)$, is equal to the supremum of cardinalities of right-separated subspaces of Z. Note that (S) is saying that $hd(X) = \aleph_0$ implies $hL(X) = \aleph_0$ for any regular space X and that (L) is the converse of this. Thus, a problem more general than (S) and (L) would ask whether $hd(X) = hL(X)$ holds for any regular space X. In this section we shall see that this strengthening of (S) and (L) is just false.

0.3. LEMMA. *If Y is either left-separated in $X[\leq]$ or right-separated in $X[\geq]$, then Y is the union of $w(X)$ conversely well-founded subposets of $X, \leq$.*

PROOF. Suppose $Y, <_w$ is a left-(right-) separated subspace of $X[\leq]$ (resp., $X[\geq]$) via separating sequence

$$\{U_x : x \in Y\}.$$

We may assume that for all x in Y there is an open B_x in X such that $U_x = B_x[\leq x]$ (resp., $U_x = B_x[\geq x]$). For open B in X, set

$$Y_B = \{x \in Y : B_x = B\}.$$

Then $x <_w y$ in Y_B implies $x \not\leq y$, so Y_B is conversely well-founded.

0.4. LEMMA. *If Y is a well-founded subposet of $X, \leq$, then Y is right-separated in $X[\leq]$ and left-separated in $X[\geq]$.*

PROOF. Since Y is well-founded with respect to $\leq$, there is a well-ordering $<_w$ of Y such that $x \leq y$ in Y implies $x \leq_w y$. It is now clear that

$$\{X[\leq x] : x \in Y\}$$

is a right-(left-) separating sequence for the well-ordering $<_w$ of Y.

The spaces $\mathbf{R}[\leq_w]$ and $\mathbf{R}[\geq_w]$ where $\leq_w$ is a well-ordering of the continuum, have been first considered by Sierpinski [45] in 1921 in order to disprove (S) and (L) in the class of Hausdorff spaces. Note that neither of the spaces $\mathbf{R}[\leq_w]$ and $\mathbf{R}[\geq_w]$ is regular and this is so for most of the quasi-orderings on $\mathbf{R}$. In this section we shall be interested in pairs $X, \leq$, where

$\leq$ is closed in X^2, i.e., the pairs for which the corresponding spaces $X[\leq]$ and $X[\geq]$ are regular. Note that this kind of a situation has been already considered in the literature but for a quite different purpose. For example if we consider $P(\omega)$ as the Cantor set via standard identification, then $P(\omega)[\subseteq]$ is the usual Vietoris topology on $P(\omega)$ ([27]). If $\leq$ is the usual ordering of the reals, then $\mathbf{R}[\leq]$ and $\mathbf{R}[\geq]$ are the usual Sorgenfrey topologies on $\mathbf{R}$ [101]. However, neither of these two examples seems relevant to (S) and (L), since for this much more care has to be made in choosing X and $\leq$ as we shall now see.

Let D be a discrete space with a fixed well-ordering $<_D$, and I be an (infinite) index-set. We shall be interested in the Tychonoff cube D^I and the following ordering on it:

$$x \leq y \text{ iff } \forall i \in I \ x_i \leq_D y_i.$$

It is clear that $\leq$ is a closed relation on D^I.

0.5. THEOREM. *If $|D| > |2^I|$ but $|I| = cf|D|$, then $hd(D^I[\leq]) < hL(D^I[\leq])$ and $hL(D^I[\geq]) < hd(D^I[\geq])$.*

PROOF. First note that $D^I, \leq$ contains no conversely well-founded subset of size bigger than 2^I. This is an immediate consequence of $|2^I|^+ \rightarrow (infinite)_1^2$. Since clearly the weight of D^I is equal to the size of D, by 0.3 and our assumption about D and I, it follows that both $hd(D^I[\leq])$ and $hL(D^I[\geq])$ are of size D. Using the cofinality assumption one can inductively construct a well-founded subposet of $D^I, \leq$ of size bigger than D. This and 0.4 gives the conclusions of 0.5.

Note that hd of any finite power of the space $D^I[\leq]$ is equal to $hd(D^I[\leq])$, and that hL of any finite power of the space $D^I[\geq]$ is equal to $hL(D^I[\geq])$. Note also that many D and I do indeed satisfy the hypothesis of Theorem 0.5. For example, we can let $I = \omega$ and $D = \omega^{th}$ successor of the continuum with the discrete topology. Note that in this case the spaces $D^I[\leq]$ and $D^I[\geq]$ are first countable. Compare all this with the Problem 8 of [75; p. 91].

Let us now consider the most interesting case when D and I are equal to the set ω of nonnegative integers, i.e., the case of Baire space ω^ω and ordering

$$a \leq b \text{ iff } \forall n < \omega \ a(n) \leq b(n).$$

It is reasonable to expect $\omega^\omega[\leq]$ and $\omega^\omega[\geq]$ to have many interesting topological and Baire categorical properties. In the rest of this section we shall be interested, however, only in certain subspaces of these spaces since they will give us some information about the conjectures (S) and (L). To begin with, let $\mathbf{P}$ be the set of all monotonically increasing elements of ω^ω. We shall also consider the following relation on ω^ω.

$$a <^* b \text{ iff } \exists m \ \forall n \geq m \ a(n) < b(n).$$

A set A of elements of $\mathbf{P}$ is *unbounded in* $\mathbf{P}$ if there is no b in $\mathbf{P}$ such that $a <^* b$ for all a in A. Note that if an A is well-ordered by $<^*$, then it is well-founded with respect to the relation $\leq$ on ω^ω. Therefore, by 0.4 it is a right-separated subspace of $\mathbf{P}[\leq]$, and it is a left-separated subspace of $\mathbf{P}[\geq]$.

0.6. THEOREM. *If A is an unbounded subset of* $\mathbf{P}$ *well-ordered by* $<^*$ *in a regular order type, then $A[\leq]$ is a right-separated first countable zero-dimensional space with no discrete subspace of size A in any of its finite powers while $A[\geq]$ is a left-separated first countable zero-dimensional space with no discrete subspace of size A in any of its finite powers.*

A subset F of some finite power A^k of A is said to be *cofinal* if

(1) $\forall a \in A \; \exists f \in F \; \forall i < k \; a <^* f_i.$

By 0.3 and 0.4, the result of 0.6 is an immediate consequence of the following combinatorial fact.

0.7. LEMMA. *If F is a cofinal subset of some finite power A^k of A, then there exist $f \neq g$ in F such that $f_i \leq g_i$ for all $i < k$.*

PROOF. Let D be a countable dense subset of F, and pick an a in A such that

(2) $\forall d \in D \; \forall i < k \; d_i <^* a.$

By the property (1) of F we can find a subset F_0 of F, still satisfying (1), and an integer m such that

(3) $\forall n \geq m \; \forall f \in F_0 \; \forall i < k \; a(n) \leq f_i(n).$

Moreover, we assume that for some s in $(\omega^m)^k$ we have:

(4) $\forall i < k \; \forall f \in F_0 \; s_i \subset f_i.$

(5) $\forall f \in F_0 \; \forall i, j < k \; (f_i <^* f_j \Rightarrow \forall n \geq m \; f_i(n) < f_j(n)).$

We may also assume that for some $\ell < k$ and for all f in F_0, f_ℓ is $<^*$ minimal among the f_i's. By looking at the minimal integer m_ℓ for which $\{f_\ell(m_\ell) : f \in F_0\}$ is unbounded in ω and then doing k successive refinements we can find integers m_i $(i < k)$, sequences t_i in ω^{m_i}, $(i < k)$, and a subset F_1 of F_0 such that

(6) $\forall f \in F_1 \; \forall i < k \; t_i \subset f_i$

(7) $\forall n < \omega \; \exists f \in F_1 \; \forall i < k \; f_i(m_i) > n.$

Pick a d in D such that

(8) $\forall i < k \; t_i \subset d_i$

and let $\bar{m}$ be an integer such that

(9) $\forall i < k \; \forall n \geq \bar{m} \; d_i(n) < a(n).$

By (7) we can pick f in F such that $f_i(m_i) > a(\bar{m})$ for all $i < k$. It is now easily checked using the monotonicity of d_i's and f_i's that $d_i(n) \leq f_i(n)$, for all $i < k$ and $n < \omega$, so we are done.

Note that a subset A of $\mathbf{P}$ satisfying the hypothesis of 0.6 can indeed be found. The statements (S) and (L) imply that it can't have the minimal possible type ω_1 under the ordering $<^*$. Thus (S) and (L) imply that every subset of ω^ω of the first uncountable size is bounded in ω^ω under the ordering $<^*$ of eventual dominance. Thus, (S) and (L) imply the negation of CH.

Fix a subset A of ω^ω. Let $\omega^\omega[A; \leq]$ be the refinement of the Baire topology on ω^ω by introducing the sets

$$\{b \in \omega^\omega : \forall n < \omega \; b(n) \leq a(n)\}, \; (a \in A)$$

as new open sets. Similarly by $\omega^\omega[A; \geq]$ we denote ω^ω with the refinement of the usual Baire topology where the sets

$$\{b \in \omega^\omega : \forall n < \omega \; a(n) \leq b(n)\}, \; (a \in A)$$

are now open sets. Note that $A[\leq]$ is equal to $A[A; \leq]$, and that $A[\geq]$ is equal to $A[A; \geq]$. Thus, for A as in 0.6, A is right-separated in $\omega^\omega[A; \leq]$ and left-separated in $\omega^\omega[A; \geq]$. Moreover, the argument of 0.6 shows that no finite power of $\omega^\omega[A; \leq]$ nor of $\omega^\omega[A; \geq]$ contains a discrete subspace of size A.

We shall now consider another subspace of $\omega^\omega[\leq]$. Fix a set A as in the hypothesis of 0.6. Furthermore we assume that

$$\forall a \in A \; \forall n < \omega \; a(n+1) - a(n) > 1.$$

For a in A let

$$\mathbf{P}_a = \{b \in \mathbf{P} : \forall n < \omega \; 0 \leq b(n) - a(n) \leq 1\}.$$

Note that $\mathbf{P}_a$ is homeomorphic to the Cantor set. Note also that each $\mathbf{P}_a$ contains a subset B_a of size continuum such that $c \not\leq b$ for all $c \neq b$ in B_a. Thus each B_a is a discrete subspace of $\omega^\omega[\leq]$. Assume now that for each a in A we can pick a subset D_a of B_a such that D_a is of smaller size than D_b whenever $a <^* b$. Hence, if we let

$$Q_A = \bigcup_{a \in A} D_a,$$

then $Q_A[\leq]$ has a discrete subspace of any size less than its own. On the other hand if a subset D of Q_A intersects D_a for cofinally many a's in A, then by 0.6 it can't be discrete. Hence $Q_A[\leq]$ contains no discrete subspace of size Q_A. To state what we have just proved, let $\mathbf{b}$ be the minimal size of an unbounded subset of ω^ω. Note that there is always an unbounded subset of $\mathbf{P}$ which has order type $\mathbf{b}$ under $<^*$.

0.8. THEOREM. *Assume $2^{\aleph_0} \geq \aleph_b$. Then the space $\omega^\omega[\leq]$ contains a right-separated subspace Q of size $\aleph_b$ which contains a discrete subspace of any size smaller than $\aleph_b$ but none of size equal to $\aleph_b$.*

Note that, this shows that the spread can be non-attained in a first countable zero-dimensional space. Clearly the similar fact can be proved for the space $\omega^\omega[\geq]$:

0.9. THEOREM. *Assume $2^{\aleph_0} \geq \aleph_b$. Then the space $\omega^\omega[\geq]$ contains a left-separated subspace Q of size $\aleph_b$ which contains a discrete subspace of any size $< \aleph_b$ but none of size equal to $\aleph_b$.*

We finish this chapter with an application of the same ideas to the well-known problem of Marczewski and Kurepa about the behavior of the cardinal function $c(X)$ (the supremum of cardinalities of disjoint families of open subsets of X) in products. The result first appeared in §1 of [59] and showed for the first time (without additional set-theoretic assumptions) that indeed the cellularity $c(X)$ is not a productive function.

0.10. THEOREM. *There is a compact zero-dimensional space X such that $c(X^2) > c(X)$.*

The proof will require a deeper analysis of the power D^I of 0.5. We start with a few definitions. Fix a regular infinite cardinal λ and a strictly increasing sequence $\{\kappa_\xi : \xi < \lambda\}$ of regular infinite cardinals. By $\prod_\xi \kappa_\xi$ we denote the set of all functions a with domain λ such that $a(\xi)$ is in κ_ξ for all $\xi < \lambda$, as well as the cardinality of this set. For a and b in $\prod_\xi \kappa_\xi$, set

$$a <^* b \text{ iff } \exists \, \eta < \lambda \, \forall \, \xi \geq \eta \; a(\xi) < b(\xi)$$

A set $A \subseteq \prod_\xi \kappa_\xi$ is *unbounded* on $I \subseteq \lambda$ if there is no b in $\prod_\xi \kappa_\xi$ such that

$$\forall \, a \in A \; \exists \, \eta < \lambda \, \forall \, \xi \in I \backslash \eta \; a(\xi) < b(\xi).$$

We shall say that A is *cofinal* in $\prod_\xi \kappa_\xi$ if for all b in $\prod_\xi \kappa_\xi$ there is an a in A such that $b <^* a$.

For a, b in $\prod_\xi \kappa_\xi$, set

$$\Gamma(a, b) = \min\{\eta < \lambda : \forall \, \xi \geq \eta \; a(\xi) \leq b(\xi)\},$$

$$\Delta(a, b) = \min\{\xi < \lambda : a(\xi) \neq b(\xi)\}.$$

When the set on the right hand side is undefined, we let the value of the function be equal to λ. In this booklet, we shall quite often work with the functions Γ and Δ but in this proof we shall only consider their restrictions on an $<^*$-increasing sequence

$$A = \{a_\alpha : \alpha < \theta\} \subseteq \prod_\xi \kappa_\xi$$

which will be fixed from now on. The cardinal θ is assumed to be regular and bigger than the supremum of κ_ξ's.

For $I \subseteq \lambda$, set

$$X_I = \{B \subseteq A : \Gamma(a, b) \in I \text{ for all } a, b \in B\}$$

$$Y_I = \{B \subseteq A : \Delta(a, b) \in I \text{ for all } a, b \in B\}.$$

(To avoid some trivial discussion, when writing $\Gamma(a, b)$ or $\Delta(a, b)$ we shall always implicitly assume that $a <^* b$.) Note that X_I and Y_I are compact subspaces of $\{0, 1\}^A$ when we identify a subset of A with its characteristic function.

0.11. LEMMA. *Suppose $\kappa_\eta > \prod_{\xi < \eta} \kappa_\xi$ for all $\eta < \lambda$. Suppose further that A is unbounded on a set $I \subseteq \lambda$ of size λ. Then the space Y_I contains no θ disjoint open sets.*

PROOF. By a standard Δ-system argument it suffices to show that for every sequence $\{F_\sigma : \sigma < \theta\}$ of finite elements of Y_I there exists $\sigma < \tau$ such that $F_\sigma \cup F_\tau$ is also a member of Y_I. First of all, note that by shrinking the sequence of F_σ's, we may assume that they are disjoint and all of the same size $n \geq 1$. Furthermore, we may assume that for some $\xi_0 < \lambda$,

(a) $\forall \sigma < \theta \ \forall a, b \in F_\sigma \ \Delta(a, b) < \xi_0$.

For $\sigma < \theta$ and $\xi < \lambda$, set

$$D_\sigma = \{\alpha < \theta : a_\alpha \in F_\sigma\},$$

$$b_\sigma(\xi) = \min\{a_\alpha(\xi) : \alpha \in D_\sigma\}.$$

Then the sequence $\{b_\sigma : \sigma < \theta\}$ is also unbounded on I, so we can find an η in I above ξ_0 such that $\{b_\sigma(\eta) : \sigma < \theta\}$ is unbounded in κ_η. Since $\kappa_\eta > \prod_{\xi < \eta} \kappa_\xi$, we can find $C \subseteq \theta$ of type κ_η and

$$\{t_0, \ldots, t_{n-1}\} \subseteq \prod_{\xi < \eta} \kappa_\xi$$

such that:

(b) $\forall \sigma < \tau$ in $C \ \forall \alpha \in D_\sigma \ \forall \beta \in D_\tau \ a_\alpha(\eta) < a_\beta(\eta)$,
(c) $\forall \sigma \in C \ \forall \alpha \in D_\sigma \ \exists i < n \ t_i \subset a_\alpha$.

It should be clear now that $F_\sigma \cup F_\tau$ is a member of Y_I for all σ and τ in C. This completes the proof.

0.12. LEMMA. *Suppose A is a cofinal subset of $\prod_\xi \kappa_\xi$ and that I is an unbounded subset of λ closed under the successor function. Then the space X_I contains no θ disjoint open sets.*

PROOF. Let $\{F_\sigma : \sigma < \theta\}$ be a given sequence of finite elements of X_I. We may again assume F_σ's are disjoint and increasing, and that for some fixed $\xi_0 < \lambda$ the condition (a) from the previous proof is satisfied for all σ.

For $\sigma < \theta$, define D_σ and b_σ as in the previous proof. Pick a function b in $\prod_\xi \kappa_\xi$ with the following property:

(d) If H is a finite subset of some κ_ξ which has the property that for some $\sigma < \theta$, H is equal to $\{a_\alpha(\xi) : \alpha \in D_\sigma\}$, then there is such a $\sigma = \sigma(H)$ with the property

$$\forall\, \eta > \xi \,\, \forall\, \alpha \in D_\sigma \,\, a_\alpha(\eta) < b(\eta).$$

(The fact that the sequence of κ_ξ is strictly increasing is used in proving that such a function b indeed exists.) Fix a $\bar\sigma < \theta$ such that $b_{\bar\sigma}$ dominates b from certain point $\xi_1 (\geq \xi_0)$ on. Moreover, we assume that every ordinal in $D_{\bar\sigma}$ dominates every ordinal appearing in $D_{\sigma(H)}$ for any H satisfying the hypothesis of (d). Pick an η in I above ξ_1 such that $\{b_\sigma(\eta) : \sigma < \theta\}$ is unbounded in κ_η. This means that there is an $H \subseteq \kappa_\eta$ satisfying the hypothesis of (d) such that every ordinal in H is bigger than any ordinal in

$$\{a_\alpha(\eta) : \alpha \in D_{\bar\sigma}\}.$$

Then by (d) and the definition of $b_{\bar\sigma}$,

(e) $\forall\, \alpha \in D_{\sigma(H)} \,\, \forall\, \beta \in D_{\bar\sigma} \,\, \Gamma(a_\alpha, a_\beta) = \eta + 1$.

Thus, $F_{\sigma(H)} \cup F_{\bar\sigma}$ is an element of X_I. This finishes the proof.

0.13. LEMMA. *If I and J are disjoint subsets of λ, then $c(X_I \times X_J) \geq \theta$ and $c(Y_I \times Y_J) \geq \theta$.*

PROOF. For a in A let $[a]$ denote the basic open set in $\{0, 1\}^A$ consisting of all functions which are equal to 1 at a. Then if we restrict

$$\{[a] \times [a] : a \in A\}$$

to any of the products we get a disjoint family of θ many nonempty open sets. This proves the Lemma.

The next Lemma finishes the proof of 0.10 because it provides situations where hypotheses of both 0.11 and 0.12 are satisfied for θ a successor cardinal.

0.14. LEMMA. *Assume $2^\lambda < \kappa$, where $\lambda = cf\kappa$. Then there is an increasing sequence $\{\kappa_\xi : \xi < \lambda\}$ of regular cardinals such that the product $\prod_\xi \kappa_\xi$ contains a $<^*$-increasing cofinal sequence of size κ^+. If, moreover, $\rho^\lambda < \kappa$ for all $\rho < \kappa$, the sequence $\{\kappa_\xi : \xi < \lambda\}$ can be chosen to have the property $\kappa_\eta > \prod_{\xi<\eta} \kappa_\xi$ for all $\eta < \lambda$.*

PROOF. Pick a sequence $\{\kappa'_\xi : \xi < \lambda\}$ of regular cardinals converging to κ and a $<^*$-increasing sequence $\{a'_\alpha : \alpha < \kappa^+\}$ in $\prod_\xi \kappa'_\xi$. Recursively on $\sigma < \bar\sigma$ define

$$\{b_\sigma : \sigma < \bar\sigma\} \subseteq \prod_\xi (\kappa'_\xi + 1)$$

such that

(f) $\forall \, \alpha < \kappa^+ \, \forall \, \sigma < \bar\sigma \ \ a'_\alpha <^* b_\sigma$

(g) $\forall \, \sigma < \tau < \bar\sigma \ \exists \, \eta < \lambda \ \forall \, \xi \geq \eta \ \ b_\tau(\xi) \leq b_\sigma(\xi)$

(h) $\forall \, \sigma < \tau < \bar\sigma \ \ |\{\xi < \lambda : b_\sigma(\xi) \neq b_\tau(\xi)\}| = \lambda.$

The ordinal $\bar\sigma$ is determined as the first place where the recursion stops. We claim that $\bar\sigma$ is a successor ordinal. Suppose $\bar\sigma$ is a limit ordinal and let us work for a contradiction. First of all note that $\bar\sigma < (2^\lambda)^+$. This is an immediate consequence of $(2^\lambda)^+ \to (\lambda^+)^2_\lambda$. For $\xi < \lambda$, set

$$B_\xi = \{b_\sigma(\xi) : \sigma < \bar\sigma\}.$$

Then the product $B = \prod_\xi B_\xi$ has size $\leq 2^\lambda$, so by the cardinality consideration there is a $\gamma < \kappa^+$ such that if for some b in B, $a'_\gamma <^* b$, then $a'_\alpha <^* b$ for all $\alpha \geq \gamma$. For $\xi < \lambda$, set

$$\bar b(\xi) = min\{\beta \in B_\xi : \beta > a'_\gamma(\xi)\}.$$

Then it is easily checked that we can put b_σ to be equal to $\bar b$ contradicting the choice of $\bar\sigma$.

Let $d = b_\sigma$, where $\sigma = \bar\sigma - 1$. For $\xi < \lambda$, let κ_ξ be the cofinality of the ordinal $d(\xi)$ and let C_ξ be a fixed club in $d(\xi)$ of order type κ_ξ. For $\xi < \lambda$ and $\alpha < \kappa^+$, set

$$a_\alpha(\xi) = tp(C_\xi \cap a'_\alpha(\xi)),$$

if $a'_\alpha(\xi) < d(\xi)$; otherwise $a_\alpha(\xi) = 0$. It is easily checked that the so constructed sequence $\{a_\xi : \xi < \kappa^+\}$ is cofinal in the product $\prod_\xi \kappa_\xi$. By going to a subsequence of κ_ξ's we may assume that they form an increasing sequence of regular infinite cardinals. If $\rho^\lambda < \kappa$ for all $\rho < \kappa$, the sequence of κ_ξ's must have κ as its supremum, so by going to a subsequence of $\{\kappa_\xi : \xi < \lambda\}$ the second conclusion of 0.14 can be satisfied. This completes the proof.

The Lemma 0.14 is given here just to indicate the applicability of Lemmas 0.11 and 0.12 and it is not really needed for the proof of 0.10. To see this assume κ is any strong limit cardinal of cofinality ω_1. Pick an increasing sequence $\{\kappa'_\xi : \xi < \omega_1\}$ of regular cardinals converging to κ and an $<^*$-increasing sequence $\{a'_\alpha : \alpha < \kappa^+\}$ in the product $\prod_\xi \kappa'_\xi$. Let d in $\prod_\xi (\kappa'_\xi + 1)$ be a $<^*$-minimal function with the property

(i) $\forall \, \alpha < \kappa^+ \ a'_\alpha <^* d.$

Using d as above, we get the sequences $\{\kappa_\xi : \xi < \omega_1\}$ and $\{a_\alpha : \alpha < \kappa^+\}$ such that the sequence of a_α's is $<^*$-increasing and $<^*$-unbounded in $\prod_\xi \kappa_\xi$. Using the fact that κ is a strong limit cardinal, it is easily seen that we can find a subsequence of κ's satisfying the hypothesis of 0.11 on which $\{a_\alpha : \alpha < \kappa^+\}$ is still unbounded. Using the fact that $\aleph_1$ is not measurable, we can find two disjoint subsets I and J of ω_1 such that $\{a_\xi : \xi < \kappa^+\}$ is unbounded on both I and J. Now Lemmas 0.11 and 0.13 give us the Theorem 0.10.

We would like to point out that the method of constructing examples of non-productiveness of the cardinal function $c(X)$ presented here is actually much more general and flexible than indicated in 0.14. For example, a slight refinement of the construction gives an example of a completely regular topological group G such that G^2 has bigger cellularity than G. Also, if the continuum is a singular cardinal, then it is a limit of cardinals with non-productive chain condition. We refer the reader to [59; §1] for further information.

0.15. Remarks.

As it has been already indicated before, some special cases of (S) and (L) have been considered in the literature since the early 1920's ([28], [45], [29]). For example, in [78] Kurepa proved that the famous Souslin Hypothesis (SH) is equivalent to the following statement, where X is an arbitrary linearly ordered space and where the collection of all open subsets of X is considered as a partially ordered set under the inclusion:

(SH) If all well-ordered chains of open subsets of X are countable, then all conversely well-ordered chains of open subsets of X are also countable.

Thus, (SH) is equivalent to (L) restricted to linearly ordered spaces. The present formulation of (S) and (L) seems to have been made in the late 1960's and early 1970's. It is not clear to us how exactly this formulation emerged in its present form, but the literature of that period does seem to give many hints to the (S) and (L) problems. For example, de Groot [17; §4] explicitly states the problem whether $d(X) \leq hc(X)$ holds for any space X. His assumptions on X seem to be only that X is Hausdorff. Thus, it seems that he was unaware of Sierpinski's [45] counterexample and, in particular, that he was unaware of the special interest of the countable case of the problem. (We have seen in this section that the answer to de Groot's problem is negative even in the class of all zero-dimensional spaces.) In [19], [76] and [77], Hajnal and Juhász (who originally started from the de Groot's paper) seem to be well aware of Sierpinski's restriction, but still don't explicitly state the problems (S) and (L). Let us also note that [39] credits the problem (S) to R. Countryman.

Most of the results about (S) and (L) obtained since the year of 1970 are stated in the forms of results about "S-spaces" and "L-spaces", i.e., counterexamples to (S) and (L), respectively. Many of the papers from that period present diagonalization arguments using forcing, CH, diamond, ... etc. to construct S- and L-spaces with various additional properties. Such a kind of results tend to be rather removed from (S) and (L) problems since, for example, the fact that CH gives S and L spaces tell us only that (S) and (L) refute CH which is generally suspected to be false anyway. The produced examples, however, have turned out to be useful to some other areas of mathematics such as measure theory (see [16]) and functional analysis (see

[79]) which have their own versions of the Ramsey problem for the uncountable. We shall actually address one such problem in Theorem 7.15 of this monograph.

The construction of this section appeared first in §4 of our paper [59] under a slightly stronger assumption on the cardinals because [50] treats some other problems under similar assumptions and we wanted to achieve a uniformity. Another place where the stronger assumption was inessential is the problem of cellularity in products considered in §1 of [59]. Namely, note that we can state Theorem 1 of [59] in the following more relaxed form

THEOREM 1'. *Assume* $\kappa > 2^{cf\kappa}$. *Then for each finite* $n \geq 1$ *there is a poset* P *such that* P^n *satisfies the* κ^+cc *but* P^{n+1} *does not.*

A detailed historical remark concerning the problem of cellularity in products can be found in [58], [59], and [84].

The space $Q_A[\leq]$ of 0.8 answers negatively the $sup = max$ problem for the cardinal function hs. This problem seems to have been first explicitly stated by Kurepa in [30; 4.2.11]. Clearly, the space $Q_A[\geq]$ of 0.9 is a counterexample to the $sup = max$ problem for the function hL, and both $Q_A[\leq]$ and $Q_A[\geq]$ will be counterexamples to the $sup = max$ problem for hc. The $sup = max$ problem for hc and hL seems to have been first asked by de Groot [17; §4]. Examples of this kind have been previously constructed by Roitman [35] (see also [22]) assuming the continuum is bigger or equal to $\aleph_{\omega_1}$, and the existence of first countable regular Luzin space. (See §6 for the definition of Luzin space.) Neither of the two constructions produce compact counterexamples to the $sup = max$ problem for hs, hL and hc. In §4 we shall state a result of [59] which produces such a counterexample using quite different methods.

Finally, we mention that the operation $X[\leq]$ can be useful in some other problems of general topology as we have shown this recently in [62n]: There is a left-separated nonmetrizable space with point-countable base which is perfectly normal and paracompact in any of its finite powers.

1. OSCILLATING REAL NUMBERS

For a and b in ω^ω let

$$osc(a,b) = |\{n < \omega : a(n) \leq b(n) \text{ but } a(n+1) > b(n+1)\}|.$$

Note that if $osc(a,b)$ is zero and if $a(0) \leq b(0)$, then $a(n) \leq b(n)$ for all n. Note also that if $osc(a,b)$ is finite, then either $a(n) \leq b(n)$ for almost all n, or $b(n) \leq a(n)$ for almost all n.

We begin our analysis of osc by introducing several useful technical definitions. First of all we shall extend osc on $(\omega^{<\omega})^2$ as follows:

For s and t in $\omega^{<\omega}$ let

$$m = min\{|s|, |t|\} - 1, \text{ and}$$

$$osc(s,t) = |\{n < m : s(n) \leq t(n) \ \& \ s(n+1) > t(n+1)\}|.$$

The letter k will always be reserved for denoting an arbitrary positive integer. When considering a k-sequence t of finite sequences of integers, i.e., a member of $(\omega^{<\omega})^k$, the ith term of t is denoted as t_i and the nth term of t_i is denoted as $t_i(n)$. If s is a finite sequence of integers of length k, let $\{s\}$ denote its range $\{s(i) : i < k\}$. Recall that $\mathbf{P}$ is the set of all increasing functions in ω^ω. For x in $\mathbf{P}^k$ and t in $(\omega^{<\omega})^k$ by $t \subset x$ we denote the fact:

$$\forall i < k \ t_i \subset x_i.$$

Note that we are using the same convention for members of $\mathbf{P}^k$: x_i is the ith member of x. For x in $\mathbf{P}^k$ by $x \mid m$ we denote the k-sequence $\langle x_o \mid m, \ldots, x_{k-1} \mid m \rangle$ of m-sequences of integers.

If $t \subset x$ are as above, then by $x[t]$ we denote the k-sequence of integers

$$\langle x_0(|t_0|), \ldots, x_{k-1}(|t_{k-1}|) \rangle.$$

If $m < \omega$, then $x[m]$ denotes the k-sequence $\langle x_0(m), \ldots, x_{k-1}(m) \rangle$. We shall say that a t in $(\omega^{<\omega})^k$ is *decreasing* if $|t_i| \geq |t_j|$ for $i < j < k$.

We shall need to consider the following relations $<_m$, $(m < \omega)$ on $\mathbf{P}^{<\omega}$:

$$x <_m y \text{ iff } \forall i < |x| \ \forall j < |y| \ \forall n \geq m \ x_i(n) < y_j(n).$$

For X and Y, two subsets of $\mathbf{P}^{<\omega}$, set

$$X <_m Y \text{ iff } \forall x \in X \ \forall y \in Y \ x <_m y, \text{ and}$$

$$X <^* Y \text{ iff } \forall \, x \in X \ \forall \, y \in Y \ x <^* y$$

Similarly for two sets M and N of integers put $M < N$ if and only if $m < n$ for all m in M and n in N. For x in $\mathbf{P}^{<\omega}$ by $\min x$ we denote the element a of $\mathbf{P}$ defined by

$$a(n) = \min\{x_i(n) : i < |x|\}.$$

We shall say that a subset X of $\mathbf{P}^k$ is *everywhere unbounded* if for all x in X and $m < \omega$, the set

$$\{y \in X : y \mid m = x \mid m\}$$

is also unbounded in $\mathbf{P}$. Clearly, every unbounded X can be refined to an everywhere unbounded subset.

For our purpose here it will be more convenient to work with the following subset of $\mathbf{P}^k$ rather than $\mathbf{P}^k$ itself:

$$[\mathbf{P}]^k = \{x \in \mathbf{P}^k : \forall \, i < j < k \ x_i <^* x_j\}.$$

For x in $[\mathbf{P}]^{<\omega}$, which is the union of $[\mathbf{P}]^k$'s, we define

$$m(x) = \min\{m : \forall \, i < j < |x| \ x_i <_{m-1} x_j\}.$$

Note that if t is equal to $x \mid m(x)$, then

$$osc(x_i, x_j) = osc(t_i, t_j) \text{ for all } i, j < |x|.$$

Now we are ready to state our main technical lemma which shows how one finds two reals with a prescribed oscillation.

1.0. LEMMA. *Suppose X and Y are two everywhere unbounded subsets of $[\mathbf{P}]^k$ and that D is a dense subset of X. Suppose further that for some decreasing t in $(\omega^{<\omega})^k$, a in $\mathbf{P}$ and $\bar{m} < \omega$, we have:*

(a) $t \subset x$ and $m(x) \le \bar{m}$ for all $x \in X \cup Y$.

(b) $D <^* a <_m Y$ for some m such that $\bar{m} < m < |t_{k-1}|$.

(c) For all $n < \omega$ there is a y in Y such that $\{y[t]\} > n$.

Then for all $\ell < \omega$ there exist d in D and y in Y such that:

(d) $osc(d_i, y_j) = osc(t_i, t_j) + \ell$ for all $i, j < k$.

PROOF. The proof is by induction on ℓ.

$\ell = 0$: Pick d in D and let m be such that $d <_m a$. By (c), pick y in Y such that

$$\{y[t]\} > a(m).$$

We claim that d and y satisfy (d) for $\ell = 0$. So pick $i, j < k$. Assume first that $i \le j$. Then by our assumptions and the choice of y it follows that:

(1) $d_i(n) = t_i(n) \le t_j(n) = y_j(n)$, for $\bar{m} \le n < |t_j|$.

(2) $d_i(n) \le d_i(m) < a(m) < y_j(|t_j|) \le y_j(n)$, for $|t_j| \le n < m$.

(3) $d_i(n) < a_i(n) < y_j(n)$, for $n \ge m$.

Thus, any n for which

(4) $d_i(n) \le y_j(n)$ but $d_i(n+1) > y_j(n+1)$.

is true, must be $\le \bar{m}$ which clearly proves (d).

Assume now $i > j$.

(5) $d_i(n) = t_i(n) > t_j(n) = y_j(n)$, for $\bar{m} \leq n < |t_i|$.

(6) $d_i(n) > d_j(n) = t_j(n) = y_j(n)$, for $|t_i| \leq n < |t_j|$.

(7) $d_i(n) \leq d_i(m) < a(m) < y_j(|t_j|) \leq y_j(n)$, for $|t_j| \leq n < m$.

(8) $d_i(n) < a(n) < y_j(n)$, for $n \geq m$.

Thus, any n for which (4) is true has to be $\leq \bar{m}$, so we are done also in this case.

$\ell + 1$: Pick $\bar{d}$ in D and $\bar{y}$ in Y such that (d) holds for $\bar{d}$, $\bar{y}$ and ℓ. Pick an integer n_0 such that:

(9) $\bar{d} <_{n_0} a$,

(10) $n_0 > |t_0|$, and

(11) $n_0 > m(\bar{d} \wedge \bar{y})$.

Let

$$Y_0 = \{y \in Y : y \mid n_0 = \bar{y} \mid n_0\}.$$

Then Y_0 is unbounded, so there is a minimal $\bar{p} \geq n_0$ such that

$$\forall n < \omega \; \exists y \in Y_0 \; \{y[\bar{p}]\} > n.$$

Hence there is an infinite subset F_0 of Y_0 such that for all y and z in F_0,

$$\{y[\bar{p}]\} < \{z[\bar{p}]\} \text{ or } \{z[\bar{p}]\} < \{y[\bar{p}]\}.$$

Thus by a sequence of k successive refinements of F_0, we may assume that there exists a decreasing u in $(\omega^{<\omega})^k$ such that:

(12) $u \subset y$ for all y in F_0.

(13) $n_0 \leq |u_{k-1}| \leq \cdots \leq |u_0| \leq \bar{p}$.

(14) For all n there is a y in F_0 such that $\{y[u]\} > n$.

Fix now $m_0 > \bar{p}$ and let

$$X_0 = \{x \in X : x \mid m_0 = \bar{d} \mid m_0\}.$$

Then X_0 is unbounded in $\mathbf{P}$, so working as above we can find a subset E_0 of X_0 and decreasing s in $(\omega^{<\omega})^k$ such that

(15) $s \subset x$ for all x in E_0

(16) $m_0 \leq |s_{k-1}| \leq \cdots \leq |s_0|$.

(17) For all $n < \omega$ there is an x in E_0 such that $\{x[s]\} > n$.

By (14) we can pick f in F_0 and e in E_0 such that

(18) $\{f[u]\} > \{e[s]\}$.

Pick $n_1 > |s_0|$ and let

$$Y_1 = \{y \in Y : y \mid n_1 = f \mid n_1\}.$$

Then Y_1 is also unbounded, so working as above we can find a subset G_1 of Y_1 and decreasing v in $(\omega^{<\omega})^k$ such that:

(19) $v \subset y$ for all y in G_1.

(20) $n_1 \leq |v_{k-1}| \leq \cdots \leq |v_0|$.

(21) For all $n < \omega$ there is a y in G_1 such that $\{y[v]\} > n$.

By (17) we can pick h in E_0 and g in G_1 such that

(22) $\{h[s]\} > \{g[v]\}$.

Pick $m_1 > |v_0|$ and pick a d in D such that

(23) $d \mid m_1 = h \mid m_1$.

Let $m_2 \geq m_1$ be such that $d <_{m_2} a$. By (21) we can find a y in G_1 such that

(24) $\{y[v]\} > a(m_2)$.

We claim that for all $i, j < k$

(25) $osc(d_i, y_j) = osc(d_i \mid n_0, y_j \mid n_0) + 1$.

This will clearly finish the proof since

$$d \mid n_0 = \bar{d} \mid n_0 \text{ and } y \mid n_0 = \bar{y} \mid n_0$$

and therefore for all $i, j < k$,

$$osc(d_i \mid n_0, y_j \mid n_0) = osc(\bar{d}_i \mid n_0, \bar{y}_j \mid n_0) =$$

$$= osc(\bar{d}_i, \bar{y}_i) = osc(t_i, t_j) + \ell$$

So fix $i, j < k$. Then by our constructions it is easily checked that:

(26) $d_i(n) = \bar{d}_i(n) < a(n) < y_j(n)$, for $n_0 - 1 \leq n < |u_j|$.

(27) $d_i(n) = s_i(n) \leq e_i(|s_i|) < f_j(|u_j|) \leq f_j(n) = y_j(n)$, for $|u_j| \leq n < |s_i|$.

(28) $d_i(n) = h_i(n) \geq h_i(|s_i|) > g_j(|v_j|) \geq v_j(n) = y_j(n)$, for $|s_i| \leq n < |v_j|$.

(29) $d_i(n) \leq d_i(m_2) < a(m_2) < y_j(|v_j|) \leq y_j(n)$, for $|v_j| \leq n < m_2$.

(30) $d_i(n) < a(n) < y_j(n)$, for $n \geq m_2$.

Hence, the only $n \geq n_0 - 1$ with property (4) is equal to $|s_i| - 1$. On the other hand, any $n < n_0 - 1$ with property (4) is counted in the computation of

$$osc(d_i \mid n_0, y_j \mid n_0).$$

This finishes our checking of (25) and also our proof of Lemma 1.0.

A subset X of **P** is *countably directed* if for any countable subset D of X there is an x in X such that $d <^* x$ for all d in D. The following result is an immediate consequence of lemma 1.0.

1.1. THEOREM. *If X is a countably directed unbounded subset of* **P**, *then for all $n < \omega$ there exist a and b in X such that $osc(a, b) = n$.*

In order to make some further applications of Lemma 1.0 we need to permute some colors of the partition osc by defining a new partition

$$o : [\mathbf{P}]^2 \to \omega$$

in the following way. We first fix an enumeration $\{\langle t^n, h^n \rangle : n < \omega\}$ of all pairs $\langle t, h \rangle$, where for some k, t is a member of $(\omega^{<\omega})^k$ and h maps $k \times k$ into ω, such that the following condition is satisfied:

(*) For all $k < \omega$, all t in $(\omega^{<\omega})^k$ and all h mapping $k \times k$ into ω, there is an $\ell < \omega$ such that for all $i, j < k$, if $n = osc(t_i, t_j) + \ell$, then $\langle t^n, h^n \rangle = \langle t, h \rangle$.

Pick $\langle a, b \rangle$ in $[\mathbf{P}]^2$. If there are no i and j such that

$$t_i^{osc(a,b)} \subset a \text{ and } t_j^{osc(a,b)} \subset b,$$

then we let $o(a, b) = 0$. Otherwise, let i and j be minimal with these properties and put

$$o(a, b) = h^{osc(a,b)}(i, j).$$

For the rest of this section, we fix an unbounded subset A of $\mathbf{P}$ which is *well-ordered* by $<^*$ in a regular order type.

1.2. THEOREM. *For any k and any cofinal subset Z of $[A]^k$ there exist a countable subset D of Z and an element b of A such that for any positive integer p, any mapping h from $k \times p$ into ω, and any x in $[A]^p$ above b there exists a d in D such that $o(d_i, x_j) = h(i, j)$ for all $i < k$ and $j < p$.*

PROOF. Clearly, we may assume Z is everywhere cofinal in A and that $m(x) = m(y)$ for all x and y in Z. Let D be a countable dense subset of Z and assume that for no b the theorem is satisfied with D and b. Hence we can find an integer p, a mapping $\bar{h}$ from $k \times p$ into ω, and a cofinal subset Y of $[A]^p$ such that

(a) $\forall d \in D \ \forall y \in Y \ \exists i < k \ \exists j < p \ o(d_i, y_j) \neq \bar{h}(i, j)$.

By shrinking Y, we may assume that for some $\bar{m} < \omega$:

(b) $\forall y \in Y \ \forall z \in Z \ m(y), m(z) \leq \bar{m}$.

Furthermore, we may assume that Y is everywhere cofinal in A. Set

$$X = \{z \ ^\wedge \ y : z \in Z, y \in Y \text{ and } z <^* y\} \subseteq [A]^{k+p}.$$

Clearly, X is cofinal in A. Fix a countable subset F of Y such that

$$E = \{d \ ^\wedge \ f : d \in D, f \in F \ d <^* f\}$$

is dense in X. Pick an a in A such that $e <^* a$ for all e in E. Working as in the proof of 0.7, we can find cofinal $W \subseteq X$ and decreasing $t \in (\omega^{<\omega})^{k+p}$ such that

(c) $t \subset x$ for all x in W.
(d) $a <_m W$ for some $\bar{m} < m < |t_{k+p-1}|$.
(e) For all $n < \omega$ there is x in W such that $\{x[t]\} > n$.

Set

$$V = \{x \in X : t \subset x\} \text{ and}$$
$$C = E \cap V.$$

Then the conditions of Lemma 1.0 are satisfied with V, W, C and t in place of X, Y, D and t. So for each $\ell < \omega$ there exist c in C and x in W such that

(f) $osc(c_i, x_j) = osc(t_i, t_j) + \ell$ for all $i, j < k + p$.

Pick any $h : (k + p) \times (k + p) \to \omega$ such that

(g) $\forall\, i < k \,\forall\, j < p\ h(i, k + j) = \bar{h}(i, j)$.

By the property (*) of the enumeration $\langle t^n, h^n \rangle$, there is an $\bar{\ell} < \omega$ such that

(h) $\forall\, i, j < k + p\ (n = osc(t_i, t_j) + \bar{\ell} \Rightarrow \langle t^n, h^n \rangle = \langle t, h \rangle)$.

Pick c in C and x in W satisfying (f) for $\ell = \bar{\ell}$. Then for some d in D, f in F, z in Z, and y in Y, we have that

(i) $c = d \wedge f$ and $x = z \wedge y$.

We claim that

(j) $\forall\, i < k \,\forall\, j < p\ o(d_i, y_j) = \bar{h}(i, j)$.

contradicting our initial assumption (a) about D, Y and $\bar{h}$. To check (j), fix $i < k$ and $j < p$. Then by (f) and (i):

(k) $osc(d_i, y_j) = osc(t_i, t_{k+j}) + \bar{\ell} = n$.

On the other hand, by (h) we have

(l) $\langle t^n, h^n \rangle = \langle t, h \rangle$.

Note that i and $k + j$ are the unique integers satisfying

(m) $t_i^{osc(d_i, y_j)} = t_i \subset d_i$, and

(n) $t_{k+j}^{osc(d_i, y_j)} = t_{k+j} \subset y_j$.

Therefore, by the definition of o and (g) it follows that

(o) $o(d_i, y_j) = h(i, k + j) = \bar{h}(i, j)$.

This checks (j) and finishes the proof of Theorem 1.2.

To see why this is relevant to the conjectures (S) and (L) let us define subspaces

$$X_A = \{x_a : a \in A\} \text{ and } Y_A = \{y_a : a \in A\}$$

of ω^A in the following way. The x_a's are determined by:

(p) $x_a(b) = 1 + o(a, b)$, for $a <^* b$,

(q) $x_a(a) = 1$,

(r) $x_a(b) = 0$, for $b <^* a$.

The y_a's are determined by:

(s) $y_a(b) = 1 + o(b, a)$, for $b <^* a$,

(t) $y_a(a) = 1$,

(u) $y_a(b) = 0$ for $a <^* b$.

Then the following facts are easy consequences of 1.2.

1.3. THEOREM. X_A *is a right-separated space of type A such that any subset X_A of size A contains a countable subset which intersects any basic open set with domain above some fixed element of A.*

1.4. THEOREM. *Y_A is a left-separated space of type A such that any family of size A of basic open sets with disjoint domains contains a countable subfamily which covers an end-segment of Y_A.*

1.5. Remarks.

In the terminology of [21], results 1.3 and 1.4 of this section show that the weak *HFD*'s and *HFC*'s do really exist if we only change the parameter ω_1 to the minimal cardinality of an $<^*$-unbounded subset of ω^ω. We believe that an analysis of this construction might be relevant to some other problems of general topology. The construction has been reproduced here from [62e] and [61] where we refer the reader for further information concerning the function *osc*. It should be noted that it is completely open whether analogues of some of the results of this chapter can also be proved for the structure $[\omega]^\omega, \subset^*$. Any such result would be of great interest.

2. THE CONJECTURE (S)
FOR COMPACT SPACES

In this section we show that the Conjecture (S) for compact spaces still has a strong influence on the reals. More precisely, we shall show that if (S) holds for compact spaces, then any subset of ω^ω of size $\aleph_1$ is bounded in ω^ω under the ordering $<^*$ of eventual dominance. Whether a similar result can be proved for the Conjecture (L) is an open problem.

From now on, we fix an unbounded subset A of $\mathbf{P}$ which has the order type ω_1 under the ordering $<^*$. For b in A, let

$$A_b = \{a \in A : a <^* b\}.$$

We also fix an $e : [A]^2 \to \omega$ such that:

(a) For all b in A, $e_b = e(\cdot, b)$ is a $1 - 1$ map from the set A_b into ω.

(b) For all a in A the set $\{e_b \mid A_a : b \in A\}$ is countable.

Thus, e is just a code for a special Aronszajn tree. For different a and b in ω^ω put

$$\Delta(a, b) = min\{n < \omega : a(n) \neq b(n)\}.$$

For notational convenience, let $\Delta(a, a) = \omega$.

We are now ready to define the main object of our study in this section, the mapping

$$H : A \to P(A)$$

defined by

$$H(b) = \{a \in A : a <^* b \ \& \ e(a, b) \leq b(\Delta(a, b))\}.$$

We shall say that an $F \subseteq [A]^k$ is *cofinal in* A, if for any a in A there is an f in F with all f_i's above a.

2.0. LEMMA. *For any finite $k \geq 1$ and cofinal $F \subseteq [A]^k$ there exist f and g in F such that $f_i \in H(g_i)$ for all $i < k$.*

PROOF. Pick a countable dense subset D of F and a c in A above any coordinate of any member of D. Since F is cofinal, we can find an integer m, an s in $(\omega^m)^k$, and cofinal $F_0 \subseteq F$ such that

(c) $\forall f \in F_0 \ \forall i < k \ s_i \subset f_i$,

23

(d) $\forall f \in F_0 \; \forall i < j < k \; \forall n \geq m \; f_i(n) < f_j(n)$,

(e) $\forall f \in F_0 \; \forall n \geq m \; c(n) < f_0(n)$.

Furthermore, we may assume that for each $i < k$ there is an $e^i : A_c \to \omega$ such that:

(f) $\forall f \in F_0 \; \forall i < k \; e^i \subseteq e_{f_i}$.

Working as in §1 we can find a t in $(\omega^{<\omega})^k$ and a subset F_1 of F_0 such that:

(g) $\forall f \in F_1 \; \forall i < k \; t_i \subset f_i$,

(h) $\forall n < \omega \; \exists f \in F_1 \; \forall i < k \; f_i(|t_i|) > n$.

Pick a d in D such that $t_i \subset d_i$ for all $i < k$, and let $\bar{m}$ be such that

(i) $\forall i < k \; \forall n \geq \bar{m} \; d_i(n) < c(n)$.

By (h) there is an f in F_1 such that

(j) $\forall i < k \; f_i(|t_i|) > max\{e^i(d_i), \, c(\bar{m})\}$.

Hence for all $i < k$

$$e(d_i, f_i) = e^i(d_i) < f_i(|t_i|) = f_i(\Delta(d_i, f_i)),$$

and so d_i is in $H(f_i)$ for all $i < k$. This completes the proof.

Note that the above proof also gives that $d_i(n) \leq f_i(n)$ for all $i < k$ and $n < \omega$ so we could have added the requirement $a \leq b$ instead of $a <^* b$ in the definition of H. But since we don't have a use for this, we keep the definition of H as it is. Note also that the proof of 2.0 shows that H has actually the following stronger property.

2.1. LEMMA. *Suppose that the Baire closure of a set $D \subseteq A^k$ contains a cofinal subset $F \subseteq A^k$. Then there exists an f in F such that for infinitely many d in D, $d_i \in H(f_i)$ for all $i < k$.*

Recursively on b in $A, <^*$ we define a subset $C(b)$ of $A_b \cup \{b\}$ containing b by putting an $a <^* b$ in $C(b)$ iff there is a c in $H(b)$ such that a is in $C(c)$ and:

(k) $\Delta(a, c) > \Delta(a, d)$ for every d in $H(b) \cup \{b\}$ different from c.

For $n < \omega$ and b in A set

$$C_n(b) = \{a \in C(b) : \Delta(a, b) \geq n\}$$

Then the following is an immediate consequence of the definition of $C(b)$ using an induction on a and b in $A, <^*$.

(l) For every a in $C_n(b)$ there is an m such that $C_m(a) \subseteq C_n(b)$.

Thus, $\{C_n(b)\}$ forms a basis for a topology on A which refines the usual Baire topology on A. Let $A[H]$ denote the associated topological space. The following is also proved by an induction.

(m) $C(b)$ is compact in $A[H]$ for all b.

Since $C(b)$ is countable, to prove this it suffices to show that every infinite $D \subseteq C(b)$ has an accumulation point in $C(b)$. If for some n,

$$D_n = \{d \in D : \Delta(d, b) = n\}$$

is infinite, then by the definition of $C(b)$ there must be c in $H(b)$ and $m < \omega$ such that $\Delta(c, b) = n$ and $C_m(c)$ is a subset of $C(b)$ having an infinite intersection with D_n. By the induction hypothesis, D_n has an accumulation point in $C_m(c)$. If no D_n if infinite, then clearly b is an accumulation point of D.

It is clear that 2.0 is saying that no finite power of $A[H]$ has an uncountable discrete subspace since clearly $H(b) \subseteq C(b)$ for all b. Thus, the one point compactification of $A[H]$ is a compact counterexample to (S). The space $A[H]$ has a few more interesting properties which we now present in some detail.

2.2. LEMMA. *For every $B \subseteq A^k$ there is an $i < k$ such that the ith projection of the difference between the Baire closure of B and the $A[H]$ closure of B is at most countable.*

PROOF. Otherwise, we could find a cofinal $F \subseteq A^k$ contained in the difference and this would easily contradict Lemma 2.1.

2.3. LEMMA. *Every finite power of the space $A[H]$ is perfectly normal.*

PROOF. The proof is by induction on the power k. Suppose $K \subseteq A^k$ is closed in $A[H]^k$. In order to show that $A[H]^k$ is perfectly normal it suffices to cover the complement of K by countably many clopen sets. For this all we need to show is that we can cover the difference $\bar{K} \setminus K$ (where $\bar{K}$ is the Baire closure of K) by countably many clopen sets of $A[H]^k$. By 2.2 some projection of $\bar{K} \setminus K$ is countable so the existence of such a covering follows easily from the induction hypothesis, the fact that $A[H]$ is locally countable, and the well-known fact (see [10; 4.5.16(b)]) that the product of a perfectly normal space and a metric space (which in our case is actually countable) is perfectly normal. (Strictly speaking we are using here the standard proof of this result (see [82; §4]) with respect to a fixed basis for the countable open subspace of $A[H]$ consisting of sets which are clopen even in $A[H]$.)

The following result summarizes what we have proved about the space $A[H]$.

2.4. THEOREM. *If there is a subset of ω^ω of size $\aleph_1$ unbounded under eventual dominance, then there is a locally countable locally compact uncountable space which is perfectly normal and hereditarily separable in any of its finite powers.*

Such a space can also be constructed assuming the existence of a Cohen real c in ω^ω. Namely in this case we define

$$H(b) = \{a \in A : a <_w b \,\&\, e(a, b) \leq c(\Delta(a, b))\}$$

where $A, <_w$ is any ground model set of reals with a well-ordering $<_w$ having type ω_1. This shows that the basic definition of H is quite flexible indeed. In fact, such an H can be obtained on any 1–1 sequence $B = \{b_\alpha : \alpha < \omega_1\}$ of members of $\{0, 1\}^\omega$ using the same tools: a $<^*$-increasing $<^*$-unbounded sequence $\{a_\alpha : \alpha < \omega_1\}$ of elements of ω^ω and a partition $e : [\omega_1]^2 \to \omega$ with properties

(a) $e_\alpha = e(\cdot, \alpha) : \alpha \to \omega$ is one-to-one for all α.

(b) $\{e_\beta | \alpha : \alpha \leq \beta < \omega_1\}$ is countable for all α.

The function $H : B \to P(B)$ is now defined as follows

$$H(b_\beta) = \{b_\alpha : \alpha < \beta \,\&\, e(\alpha, \beta) \leq a_\beta(\Delta(b_\alpha, b_\beta))\}.$$

Then as before $H(b_\beta)$ is either finite or it is a sequence which converges to b_β. A proof almost identical to that of 2.0 shows that the new H also satisfies 2.0 (and 2.1), so the associated space $B[H]$ satisfies 2.4 as well. Thus we have proved the following.

2.5. THEOREM. *If there is a subset of ω^ω of size $\aleph_1$ unbounded under eventual dominance, then the topology of any set of reals of size $\aleph_1$ can be refined to a locally countable locally compact topology which is hereditarily separable and perfectly normal in all of its finite powers.*

2.6. Remarks.

The result of this section first appeared in our note [62d]. The first compact counterexamples to (S) have been constructed by Fedorchuk [14] and Ostaszewski [33] via diagonalization arguments which used the Jensen diamond principle. A compact counterexample to (S) via a CH-diagonalization argument was constructed, since then, by Kunen ([24], [23]). Let us mention that CH also gives a compact counterexample to (L) which is another result of Kunen ([80]).

The first counterexamples to (S) and (L) in the one Cohen real extension were given by Roitman [36].

3. SOME PROBLEMS CLOSELY RELATED TO (S) AND (L)

In the first three sections we have seen that many examples of spaces, obtained in the literature in very special models of set theory, can, in fact, be made alive using constructions not going beyond the usual axioms of set theory. We believe that this is a promising line of research not only in attacking (S) and (L) but also many other problems of general topology. The purpose of this section is to illustrate this on a few problems closely related to (S) and (L).

Fix an $<^*$-unbounded subset A of P such that $A, <^*$ is a well-ordered set having a regular order type. Let $<_{lex}$ be the usual lexicographical ordering of ω^ω. As it has been already noted $A[\leq_{lex}]$ and $A[\geq_{lex}]$ are the usual Sorgenfrey topologies on A. It should be clear that the combinatorial fact 0.7 has the following immediate corollary.

3.0. THEOREM. *Neither $A[\leq_{lex}]$ nor $A[\geq_{lex}]$ have left nor right separated subspace of size A in any of their finite powers. But, clearly, their product contains a closed discrete subspace of size A.*

Concerning this example we have the following curiosity.

3.1. THEOREM. *There exist first countable Hausdorff spaces X and Y which are hereditarily separable and hereditarily Lindelöf in any of their finite powers but $X \times Y$ contains an uncountable discrete subset.*

PROOF. Pick a pair $S, \leq_s$ where S is an uncountable set of reals and $\leq_s$ is a total ordering of S such that any finite power S^k with the product ordering $\leq_s^k$ is the union of countably many chains (see [43] and [60]). Then the Lemmas of §0 are sufficient for concluding that $X = S[\leq_s]$ and $Y = S[\geq_s]$ are as required.

The spaces X and Y contain no uncountable regular subspaces. This follows from the following general fact which gives us a hint that the choices of the orderings in §§0 and 1 are, in a sense, unavoidable.

27

3.2. PROPOSITION. *Let S be a set of reals and let $<_s$ be a total ordering of S such that each $S[\leq_s]$ is a regular space. Then S can be decomposed into countably many sets S_n such that for all n and x in S_n the set $\{y \in S_n : y \leq_s x\}$ is closed in S_n.*

PROOF. For any x there is a rational interval I_x containing x such that the closure of $\{y \in I_x : y \leq_s x\}$ in $S[\leq_s]$ is disjoint from the closed set $\{y \in S : x <_s y\}$. The required decomposition consists of sets $S_I = \{x \in S : I_x = I\}$, where I is a rational interval.

Hence, any total ordering $<_s$ on S for which $S[<_s]$ is regular is the union of countably many suborderings each of which is isomorphic to a set of reals. This difficulty in finding "right" orderings on sets of reals might well be connected with the following Conjecture closely related to (S) and (L).

(M) *If X is a regular space and if X^2 contains no uncountable discrete subspace, then $X^{\aleph_0}$ is hereditarily Lindelöf and hereditarily separable, and maybe we can even conclude that X has a countable network (i.e., a countable collection $\mathcal{N}$ such that for every x in X and open U containing x there is an N in $\mathcal{N}$ such that $x \in N \subseteq U$).*

The truth and consistency of (M) are still widely open problems. The conjecture (M) has similar influence on real numbers as (S) and (L). This can be shown by looking at certain spaces generated by the set A of 0.6 which has the following property when Baire space is identified with the irrationals using continued fractions.

3.3. THEOREM. *For any finite k and any open set U in $\mathbf{R}^k$ containing $\mathbf{Q}^k$ there is an a in A such that any k-tuple of A above a is in U.*

PROOF. For a finite or infinite sequence $s_0, s_1, \ldots$ of elements of ω let $[s_0, s_1, \ldots]$ denote the continued fraction which will be identified with the corresponding real number (see [74]). The rationals have finite continued fractions $[s_0, s_1, \ldots, s_{n-1}]$. Note that

$$\{[s_0, s_1, \ldots, s_{n-1}, i] : i < \omega\}$$

converges to $[s_0, s_1, \ldots, s_{n-1}]$ from the right. Suppose we have open U in $\mathbf{R}^k$ containing $\mathbf{Q}^k$ but there is a cofinal $F \subseteq A^k$ such that the set of all continued fractions of elements of F is disjoint from U. Working as in the proof of 0.7 we can find m_i $(i < k)$, $t_i \in \omega^{m_i}$ $(i < k)$ and $F_1 \subseteq F$ such that

(a) $\forall f \in F_1 \; \forall i < k \; t_i \subset f_i$
(b) $\forall n < \omega \; \exists f \in F_1 \; \forall i < k \; f_i(m_i) > n$.

This means that the element

$$\langle [t_i(0), \ldots, t_i(m_i - 1)] : i < k \rangle$$

of $\mathbf{Q}^k$ is a limit of the set of continued fractions determined by F_1, a contradiction.

The result of 3.3 is relevant to the constructions of Michael [32] because an easy induction on k shows that A of 3.3 has the following property

(MP) For every open U in $\mathbf{R}^k$ containing $(A \cup \mathbf{Q})^k \setminus A^k$ the difference $A^k \setminus U$ has size smaller than A.

An uncountable set A with this property has been first constructed in [32] via a classical diagonalization argument which heavily uses CH. In [32] Michael uses such a kind of sets in producing counterexamples to (M) and similar problems. In 3.6 below we shall present another construction giving a counterexample to (M) assuming A has type ω_1 under $<^*$.

There are many statements in the literature closely related to (M) and at the moment it is not clear which of them are more basic than the others. One of them is related to the following concept which naturally generalizes the concept of right and left separation considered before. We say that a subset S of a space X is *weakly-separated* if for every s in S we can choose open U_s containing s such that if s and t are different members of S, then either s is not in U_t or t is not in U_s. Clearly, a space with an uncountable weakly separated subset can't have a countable network. Whether the converse of this implication is true is the content of the following statement.

(T) If X is a regular space with no uncountable weakly separated subspace, then X has a countable network.

Let T_X be the set of all pairs $\langle x, U \rangle$ where x is a point of X and U is an open set containing x. To X we associate a partition

$$[T_X]^2 = K_X^0 \cup K_X^1$$

defined by: $\{\langle x, U \rangle, \langle y, V \rangle\}$ is in K_X^0 iff either x is not an element of V or else y is not an element of U. Then the statement (T) is equivalent to the Ramsey-type statement asserting that for every regular space X either T_X contains an uncountable 0-homogeneous set for the associated partition or else T_X is the union of countably many 1-homogeneous sets. In §8 we shall present such a statement for a special class of regular spaces X. If X does have an uncountable weakly separated subspace one might wish to consider the anti-symmetric relation which gives the weak separation. The following statement says, among other things, that there exist essentially only two such relations, the ordering of the reals and the equality relation.

(G) If X is a regular space, then X has either an uncountable discrete subspace, or a countable network, or an uncountable subspace of the Sorgenfrey line.

Clearly, (T) is a part of (G) since a subspace of Sorgenfrey line is weakly separated. Assuming that a square of an uncountable subspace of the Sorgenfrey line has an uncountable discrete subspace (a situation which can easily be arranged; see 8.4(c)), the Conjecture (M) is also a consequence of (G).

In the rest of this section we shall construct some examples showing strong influence of (M), (T) and (G) on the reals. To state our first result, let us consider the following weakening of both (T) and (G).

(W) If X is a regular space such that no finite power of X contains an uncountable weakly separated subspace, then X has a countable network.

The methods of §8 can be used to show that (W) is a consequence of the Proper Forcing Axiom considered there and that, in fact, (W) is much weaker than (T).

3.4. THEOREM. *If (W) holds, then the minimal cardinality of an $<^*$-unbounded subset of ω^ω is equal to $\aleph_2$.*

In proving 3.4 we shall present two constructions of independent interest depending whether the minimal cardinality **b** of an $<^*$-unbounded subset of ω^ω is equal to $\aleph_1$ or is bigger than $\aleph_2$.

Pick an $<^*$-unbounded subset A of ω^ω such that the order type of $A, <^*$ is a regular cardinal. Our first construction uses the function $o : [A]^2 \to \omega$ of §1. To simplify the notation, let $o\{x, y\}$ denote either $o(x, y)$ or $o(y, x)$ depending whether $x <^* y$ or $y <^* x$, respectively. We shall also agree that $o\{x, x\}$ is equal to 0 for all x in A. For x in A, set

$$U_x^0 = \{y \in A : o\{x, y\} = 0\}.$$

Let U_x^1 be the complement of U_x^0 in A. Let $A[o]$ denote the set A with the topology generated by

$$\{U_x^i : x \in A, i < 2\}$$

as a subasis. Thus, a typical basic open set is determined by a function σ from a finite subset D of A into 2 as follows

$$[\sigma] = \bigcap\{U_x^{\sigma(x)} : x \in D\}.$$

Then $A[o]$ is a zero-dimensional space. An easy consequence of 1.2 is that an end-segment of A is a Hausdorff (and therefore completely regular) subspace of $A[o]$.

3.5. THEOREM. *Every weakly separated subspace of some finite power of $A[o]$ has fewer elements than A. Every network of $A[o]$ has size A.*

PROOF. The second sentence follows easily from the fact that for every cofinal subset B of A there exists $x \neq y$ in B such that $o\{x, y\} \neq 0$ i.e., x is not in U_y^0 and y is not in U_x^0. This fact is obviously a consequence of 1.2. So let us concentrate on the first conclusion. We shall prove it only for the case of $A[o]$ itself since the general case is only notationally different. Suppose we are given cofinal subset B of A and for each x in B a function σ_x from a finite subset D_x of A containing x such that x is in $[\sigma_x]$. We have to find $x \neq y$ in B such that x is in $[\sigma_y]$ and also y is in $[\sigma_x]$. By a Δ-system argument we may assume that the sets D_x are disjoint and of the same size k. Moreover, we may assume that for some $l < k$ and σ in $\{0, 1\}^k$ the following is true for every x in B:

(c) x is the lth element of D_x,
(d)   If y is the ith element of D_x, then $\sigma_x(y) = \sigma(i)$.

Now, to get the required x and y in B, we apply 1.2 to a function $h : k \times k \to 2$ such that

$$h(l, i) = h(i, l) = \sigma(i) \quad \text{for all } i < k.$$

Note that if we replace "$= 0$" by "$\neq 0$" in the above construction the associated space will also satisfy 3.5 but the product of these two spaces will have the diagonal as a closed discrete subspace of size A. An obvious generalization of this construction gives us the following result which shows that the statement (M) also implies that every subset of ω^ω of size $\aleph_1$ is bounded.

3.6. THEOREM. *For every finite $n \geq 1$ there is a zero-dimensional space X such that X^n contains no weakly separated subspace of size $\mathbf{b}$ but X^{n+1} has a closed discrete subspace of size $\mathbf{b}$.*

Note that 3.5 proves 3.4 in the case of $\mathbf{b} = \aleph_1$. The next result takes care of the remaining case. Recall that a space X is said to be *cometrizable* if there is a weaker metric topology on X with property that each point of X has a neighborhood base consisting of sets which are closed in the metric topology.

3.7. THEOREM. *If every subset of ω^ω of size at most $\aleph_2$ is bounded, then there is a cometrizable zero-dimensional space X of size $\aleph_2$ such that X has no countable network but any smaller subspace of X does have a countable network.*

The proof of 3.7 will use a method of naturally associating topological spaces with gaps in ω^ω. This construction can also be considered as a variation on $\omega^\omega[\leq]$ of §0. A *gap* in $\omega^\omega, <^*$ is a subset of ω^ω of the form $A \cup B$ such that:

(1) $tp(A, <^*)$ is a regular infinite cardinal;
(2) $tp(B, <^*)$ is the converse of a regular infinite cardinal;
(3) $f <^* g$ for all f in A and g in B;
(4)   there is no h in ω^ω such that $f <^* h <^* g$ for all f in A and g in B.

Fix a gap $\langle A, B \rangle$ in $\omega^\omega, <^*$ with both A and B uncountable. For $f <^* g$ in ω^ω, set

$$\Gamma(f, g) = min\{m : f(n) < g(n) \text{ for all } n \geq m\}.$$

The space associated with $\langle A, B\rangle$ will be generated by sets

$$U_{ng}(f) = \{h \in A : \Delta(f, h) \geq n \geq \Gamma(h, g)\},$$

where $\Gamma(f, g) \leq n < \omega$, $f \in A$, $g \in B$. Note that each $U_{n,g}(f)$ is a closed set in the Baire topology on A. Thus $\{U_{ng}(f)\}$ forms a basis for a zero-dimensional cometrizable refinement of the usual Baire topology on A. The new topological space will be denoted henceforth by $A\langle B\rangle$.

3.8. LEMMA. *The space $A\langle B\rangle$ does not have a network of size smaller than A and B.*

PROOF. Otherwise, there would be a decomposition $\{W_i\}_{i \in I}$ of $A \times B$ such that $|I| < min\{|A|, |B|\}$ and such that for every i in I there is an $n_i < \omega$ such that:

(5) $\Gamma(f, g) = n_i$ for all $\langle f, g\rangle$ in W_i,
(6) if $\langle f, g\rangle$ and $\langle \bar{f}, \bar{g}\rangle$ are in W_i, then $\Gamma(f, \bar{g}) \leq n_i$ and $\Gamma(\bar{f}, g) \leq n_i$.

By the regularity of $A, <^*$ and $B, >^*$ there exist j in I and cofinal $\bar{A} \subseteq A$ such that for every f in $\bar{A}$ there is cofinal $B_f \subseteq B$ such that $\langle f, g\rangle$ is in W_j for all f in $\bar{A}$ and g in B_f. Let $\bar{f}$ be the minimum of $\bar{A}$ and define h in ω^ω by letting $h \mid n_j$ to be arbitrary but for $n \geq n_j$,

$$h(n) = min\{g(n) : g \in B_{\bar{f}}\}.$$

It should be now clear from (5) and (6) that h splits the gap $\langle A, B\rangle$, a contradiction.

3.9. LEMMA. *Every smaller subspace of $A\langle B\rangle$ is the union of countably many metric subspaces, and so, in particular, it has a countable network.*

PROOF. Let $\bar{A} \subseteq A$ have size smaller than A, and let $\bar{f}$ in A be an upper bound of $\bar{A}$. For $n < \omega$ let

$$\bar{A}_n = \{f \in \bar{A} : \Gamma(f, \bar{f}) \leq n\}.$$

Then it follows easily from the definition of the space $A\langle B\rangle$ that the $A\langle B\rangle$-topology and the Baire topology agree on $\bar{A}_n$ for all n. This finishes the proof.

Note that the proof of 3.9 also shows that the weight of the space $A\langle B\rangle$ is equal to the size of B. To finish the proof of 3.7 all we have to prove is the following.

3.10. LEMMA. *If every subset of ω^ω of size $\leq \aleph_2$ is $<^*$-bounded, then there is a gap $\langle A, B\rangle$ in $\omega^\omega, <^*$ such that the type of $A, <^*$ is equal to ω_2 and the type of $B, >^*$ is a regular uncountable cardinal.*

PROOF. Pick a subset A of ω^ω which has type ω_2 under $<^*$ and extend it to a $\subseteq$-maximal $<^*$-totally ordered subset L of ω^ω. The set A will determine a gap in L whose coinitiality cannot be 0, 1 or ω by a classical result of Rothberger (see [38] and [8]), and this clearly finishes the proof.

The gap spaces can also be considered in the structure $[\omega]^{\omega}, \subset^*$ rather than $\omega^{\omega}, <^*$ using the obvious translation. Note that $\omega^{\omega}, <^*$ and $[\omega]^{\omega}, \subset^*$ have the same types of gaps, so for our purposes here it doesn't make any difference which of the two structures we consider.

3.11. Remarks.

The Hausdorff spaces constructed in 3.1 first appeared in our note [62c].

The first counterexamples to the Conjecture (M) using CH-diagonalization arguments appeared in Michalel's paper [32]. See also [85] and [86] for stronger CH-counterexamples.

Concerning the property (MP) and Rothberger's equivalence of the existence of an uncountable set of reals concentrated around **Q** (see [38] and [8]), we mention the following consequence of 3.3:

> *If there is an uncountable set of reals concentrated around* **Q**, *then there is such a set which moreover has the property C″ (see [87]) in all of its finite powers.*

The statement (T) seems to have been first considered by Tkacenko [51]. In [18] Gruenhage considers the statement (G) and proves its consistency in the class of cometrizable spaces by essentially deducing it from the Open Coloring Axiom for sets of reals which will be considered here in §8 (see 8.5).

The results 3.4-3.10 first appeared in our note [62m] where we refer the reader for a few more remarks. The statement (T) has been also considered by Hajnal and Juhász [88] and Ciesielski [89] in some special models of set theory.

Concerning Martin's axiom and the existence of (κ, λ^*) gaps (and, therefore, the existence of corresponding gap-spaces of 3.7), we mention the following facts from Kunen [25] (see also [4] or §§7 and 8). If MA holds and there is a (κ, λ^*) gap in ω^{ω}, then either one of the κ or λ is equal to the continuum or else $\kappa = \lambda = \omega_1$ (see §7). Thus, for every uncountable $\kappa < 2^{\aleph_0}$ different from ω_1 we can construct a $(\kappa, (2^{\aleph_0})^*)$ gap. The gaps with types (κ, λ^*) for $\kappa, \lambda \in \{\omega_1, 2^{\aleph_0}\}$ can exist in a model of MA but their existence doesn't follow from MA (see §8).

4. DIAGONALISATIONS
OF LENGTH CONTINUUM.

In this section we present several applications and generalizations of the classical diagonalization argument on the continuum. We first mention the following results from [59].

4.0. THEOREM. *There is a compact zero-dimensional ccc space X such that the continuum is not a precaliber of X.*

(That is, there is a family of $2^{\aleph_0}$ clopen sets in X which contains no subfamily of size $2^{\aleph_0}$ with the finite intersection property.) The space of 4.0 will be explained in §7 and is obtained by naturally associating a partial ordering to an object obtained by a classical diagonalization argument. This is also the case with the following space of [59; §3] which we state here in a slightly more explicit form.

4.1. THEOREM. *There is a compact first countable zero-dimensional space X of size continuum such that*

(a) *X has a discrete subspace of any size smaller than the continuum but none of the size equal to continuum.*

(b) *If Y is a closed subspace of X, then Y has a basis $\mathcal{B}$ which can be decomposed as $\mathcal{A} \cup \bigcup_{i<\omega} \mathcal{B}_i$, where $\mathcal{A}$ is the set of singletons which are isolated in Y and where for each $i < \omega$, $\mathcal{B}_i$ has the finite intersection property.*

Thus, if the continuum is a limit cardinal, the spread (i.e., hc) of the space X is not attained. The part (b) of 4.1 is saying that every closed subspace of X has cellularity less than the continuum. (By the way, the space X happens to be separable.) So, we have here a compact counterexample to the $sup = max$ problem for the cardinal function spread which is quite different from the one obtained in the constructible universe using a Souslin tree at an inaccessible cardinal κ (see [83]). In the κ-Souslin tree example both cellularity and the spread are equal to κ. Moreover, provided the continuum is a singular cardinal, 4.1 produces the first known compact counterexample to the $sup = max$ problem for hc at a singular level.

One of the basic tools of the classical diagonalization argument is the notion of oscillation of an

$$f : A \to \mathbf{R}$$

at some point x of $\mathbf{R}^n$ (where $A \subseteq \mathbf{R}^n$) defined as follows (see [27]):

$$\omega_f(x) = \bigcap \{\overline{f''(A \cap I)} : I \text{ open in } \mathbf{R}^n \text{ and } x \in I\}.$$

(The closure is taken in the extended line $\mathbf{R} \cup \{\pm\infty\}$.) This allows us to consider the set A_f^* of all x in $\mathbf{R}^n$ with $|\omega_f(x)| = 1$ on which the function f determines a continuous function f^*. Note that A_f^* is a G_δ subset of $\mathbf{R}^n$. By enumerating all such A_f^*'s (and the way f's act on them) and then diagonalizing over such an enumeration we are able to produce many interesting examples. Some such examples will be produced in this section because of their use later on. But first we would like to give the example 4.1 because it uses the simplest kind of a diagonalization argument–a diagonalization argument in the dimension one.

PROOF OF 4.1. Let $\{f_\xi : \xi < 2^{\aleph_0}\}$ enumerate all continuous functions from G_δ subsets of $\mathbf{R}$ into $\mathbf{R}$. Recursively on $\alpha < 2^{\aleph_0}$ we shall construct disjoint subsets A_α of $\mathbf{R}$ such that the cardinality of A_α is equal to $|\alpha| + \aleph_0$. Let A_0 and A_1 be two disjoint countable dense subsets of $\mathbf{R}$. Suppose we have A_α for all $\alpha < \beta$, ($\beta \geq 2$). Let

$$B_\beta = \bigcup_{\alpha < \beta} A_\alpha,$$

and let B_β^* be the minimal subset of $\mathbf{R}$ which contains B_β and which is closed under any of the functions f_ξ for $\xi < \beta$. Then B_β^* has size $|\beta| + \aleph_0$, so we can find t_β in $\mathbf{R}$ such that

$$A_\beta = \{t_\beta + r : r \in B_\beta\}$$

is disjoint from B_β^*. This completes the recursive definition of A_α's.
      Let

$$E_0 = \bigcup \{A_\alpha : \alpha < 2^{\aleph_0}, \alpha \text{ even}\}, \text{ and}$$

$$E_1 = \bigcup \{A_\alpha : \alpha < 2^{\aleph_0}, \alpha \text{ odd}\}.$$

Then E_0 and E_1 are disjoint dense subsets of $\mathbf{R}$ with property that for any $\kappa < 2^{\aleph_0}$ there is an increasing function from a subset of E_0 of size κ into E_1. We claim that there is no monotonic (increasing or decreasing) function f from a subset A of E_0 of size $2^{\aleph_0}$ into E_1. For x in $E_0 \cup E_1$, let $\alpha(x)$ be the unique α such that x is in A_α. By considering the inverse of f, we may assume that $y = f(x)$ implies $\alpha(x) < \alpha(y)$. By removing a countable subset of A, we may assume f is continuous on A. Let A_f^* and f^* be determined by A and f as in the above remark concerning the definition of the oscillation function. Then A_f^* contains A and there is a ξ such that $f_\xi = f^*$. Pick an x in A such that $\alpha(x) > \xi$ and let $y = f(x)$. Then $x \in B_{\alpha(y)}$ implies $y = f(x) = f_\xi(x)$ is in $B_{\alpha(y)}^*$ which is disjoint from $A_{\alpha(y)}$, a contradiction.

Now, let us return to the actual construction of the space X of 4.1. By using an isomorphism between the interval $(0,1)$ and $\mathbf{R}$, we may assume that E_0 and E_1 are subsets of $(0,1)$. For $i < 2$, let K_i be the linearly ordered set obtained from $[0,1]$ by replacing every r in E_i by two new points $r^0 < r^1$. We claim that $X = K_0 \times K_1$ satisfies 4.1.

First of all, note that if f is an increasing mapping from a subset A of E_0 into E_1, then

$$\{\langle r^0, f(r)^1 \rangle : r \in A\}.$$

is a discrete subspace of X. So X contains a discrete subspace of any size $< 2^{\aleph_0}$.

Suppose D is a discrete subspace of X of size $2^{\aleph_0}$. Since no K_i contains an uncountable discrete subspace, we may assume that D is a graph of a 1–1 function from a subset of K_0 into K_1. Furthermore, we may assume that for some fixed $i, j < 2$ every member of D has the form $\langle r^i, s^j \rangle$ for some r in E_0 and s in E_1. We may also assume that the separating neighborhoods of elements of D are products of intervals of E_0 and E_1, and by shrinking D, we may assume that the intervals of E_0 contain a fixed rational q_0 and the intervals of E_1 contain a fixed rational q_1. Now it is easily checked that D induces an increasing or decreasing function from a subset of E_0 of size $2^{\aleph_0}$ into E_1 depending on whether i and j are different or equal. This contradicts our choice of E_0 and E_1.

To show that X satisfies 4.1(b) fix a closed subspace Y of X. Clearly, Y has a basis consisting of (restrictions to Y of) products of intervals of K_0 and K_1. Whenever we can, we shall consider only the intervals whose both or at least one end-point is a rational number. This splits such products into countably many groups depending on types of intervals involved. Hence it suffices to decompose every such group into subfamilies as claimed in 4.1(b). To illustrate this, suppose we have a rectangle of the form

$$P = [x, p) \times [y, q),$$

where $\langle x, y \rangle$ is an element of Y such that for some r in E_0 and s in E_1, $x = r^1$ and $y = s^1$, and where p and q are rationals. If $\langle x, y \rangle$ is an isolated point of Y, we replace the rectangle with a smaller rectangle separating this point from the rest of Y. So suppose $\langle x, y \rangle$ is not isolated and pick $\langle u, v \rangle$ in $Y \cap P$ different from $\langle x, y \rangle$.

If $u = x$, let $i(P) = *$ and let $j(P)$ be any rational such that

$$y < j(P) < v.$$

If $v = y$, let $j(P) = *$ and let $i(P)$ be any rational such that

$$x < i(P) < u.$$

If $u \neq x$ and $v \neq y$ let $i(P)$ and $j(P)$ be any two rationals such that

$$x < i(P) < u \text{ and } y < j(P) < v.$$

It is clear that the family of all rectangles (of the same type as P) which have the same pair of parameters is centered when restricted to Y. This completes our proof of 4.1.

4.2. THEOREM. *If X and Y are sets of reals of size continuum, then there exists a 1-1 map $f : X \to Y$ such that every disjoint family of $2^{\aleph_0}$ finite increasing subfunctions of f has two members whose union is also increasing.*

PROOF. Let $\{x_\xi : \xi < 2^{\aleph_0}\}$ be a 1-1 enumeration of X and let $\{g_\xi : \xi < 2^{\aleph_0}\}$ enumerate all continuous functions from G_δ subsets of some finite power of $\mathbf{R}$ into $\mathbf{R}$. Let $f(x_\alpha)$ be any element of

$$Y \setminus \{f(x_\xi) : \xi < \alpha\}$$

which is not of the form $g_\xi(p^\wedge x_\alpha)$, where $\xi < \alpha$ and p is a finite sequence of elements of

$$\{x_\xi : \xi < \alpha\} \cup \{f(x_\xi) : \xi < \alpha\}.$$

We claim that f works. So pick a disjoint family K of size $2^{\aleph_0}$ of finite increasing subfunctions of f. We may assume that for some $n \geq 1$, $|s| = n$ for all s in K. The proof is by induction on that n, and is essentially the same as that of [58; §1], but for the convenience of the reader we give the argument. To K we associate the following function g from a subset of $\mathbf{R}^{2n-1}$ into $\mathbf{R}$: If

$$s = \{\langle x_{\alpha_i}, f(x_{\alpha_i}) \rangle : i < n\},$$

where α_i's are increasing, is an element of K, then

$$\bar{s} = \langle x_{\alpha_0}, f(x_{\alpha_0}), \dots, x_{\alpha_{n-1}} \rangle$$

is a typical member of $\mathrm{dom}(g)$ and

$$g(\bar{s}) = f(x_{\alpha_{n-1}}).$$

By shrinking K we may assume that some fixed set of $2n$ rational intervals separates every element of K. By our choice of f it follows that the set K_0 of all elements s of K for which $|\omega_g(\bar{s})| \geq 2$ is of size continuum. By shrinking K_0 and by symmetry, we may assume that for some rational d and all s in K_0 there is a x in $\omega_g(\bar{s})$ such that $g(\bar{s}) < d < x$. By the inductive hypothesis there exist $s \neq t$ in K_0 such that the first $n-1$ elements of s and the first $n-1$ elements of t form an increasing function. By symmetry assume that $x < y$, where $\langle x, f(x) \rangle$ is the last element of s and $\langle y, f(y) \rangle$ is the last element of t. Let I be the product of $2n - 1$ rational intervals such that $\bar{t} \in I$ and no coordinate of I contain a member of s. By the property of K_0 there is an u in K such that $\bar{u} \in I$ and $g(\bar{u}) > d$. Then $s \cup u$ is increasing.

4.3. Remark. Note that the proof of 4.2 shows that the poset P of all finite increasing subfunctions of f actually has the following slightly stronger property: If $K \subseteq P^{<\omega}$ has size continuum and has the property that $p \neq q$ in

K implies $p_i \cap q_j = \varnothing$ for all i and j. Then there exist $s \neq t$ in K such that $s_i \cup t_i$ is increasing for all i.

In the next result we shall find it more convenient to work with the Cantor space $\mathbf{C} = \{0, 1\}^\omega$. By $\exp(\mathbf{C})$ we denote the set of all nonempty closed subsets of $\mathbf{C}$ with the standard exponential topology where the nth basic open neighborhood of an F in $\exp(\mathbf{C})$, $B_n(F)$, is equal to

$$\{E \in \exp(\mathbf{C}) : T_E \cap 2^n = T_F \cap 2^n\},$$

and where for F in $\exp(\mathbf{C})$, T_F is the tree determined by F, i.e.,

$$T_F = \{s \in \{0, 1\}^{<\omega} : s \subset f \text{ for some } f \text{ in } F\}.$$

It is well-known, and easily checked, that $\exp(\mathbf{C})$ is homeomorphic to $\mathbf{C}$ (see [27]).

A *closed set mapping* is a function F from a subset of $\mathbf{C}$ into $\exp(\mathbf{C})$. A set $X \subseteq \mathrm{dom}(F)$ is *F-free* iff

$$F(x) \cap X \subseteq \{x\} \text{ for all } x \text{ in } X.$$

A set $X \subseteq \mathrm{dom}(F)$ is F-*connected* if for all $x \neq y$ in X either $x \in F(y)$ or $y \in F(x)$.

4.4. THEOREM. *If F is a closed set mapping with domain a subset of the Cantor space which is not the union of $< 2^{\aleph_0}$ connected subfunctions, then there is a set $Y \subseteq \mathrm{dom}(F)$ of size $2^{\aleph_0}$ such that every disjoint family of $2^{\aleph_0}$ many finite F-free subsets of Y has two members whose union is also F-free.*

PROOF. Let $X = \mathbf{C} \times \exp(\mathbf{C})$. We shall identify F with the subset $\{\langle x, F(x) \rangle : x \in \mathrm{dom}(F)\}$ of X. For p in X^n and open $U \subseteq X^n$ containing p let

$$U_p = \{q \in U : q_i \neq p_i \text{ and } \{p_i, q_i\} \text{ is free for all } i < n\}.$$

If f is a function from a subset A of X^n into X and p is in X^n, then

$$\omega_f^*(p) = \bigcap\{\overline{f''(U_p \cap A)} : U \subseteq X^n \text{ is open and } p \in U\}.$$

Let $\{f_\xi : \xi < 2^{\aleph_0}\}$ enumerate all countable functions from a finite power of X into X and let $\{T_\xi : \xi < 2^{\aleph_0}\}$ enumerates all closed connected subfunctions of F. The set Y will be equal to $\{x_\xi : \xi < 2^{\aleph_0}\}$ where:

(a) x_α is an element of $\mathrm{dom}(F) \setminus \{x_\xi : \xi < \alpha\}$,
(b) $x_\alpha \notin \mathrm{dom}(T_\xi)$ for all $\xi < \alpha$, and
(c) $\langle x_\alpha, F(x_\alpha) \rangle$ is not a member of any connected function of the form $\omega_{f_\xi}^*(p) \cap F$, where $\xi < \alpha$ and where p a finite sequence of elements of

$$\{\langle x_\xi, F(x_\xi) \rangle : \xi < \alpha\}.$$

Let K be a disjoint family of $2^{\aleph_0}$ finite free subfunctions of $F \mid Y$. We may assume that all elements of K have the same size $n \geq 1$. The proof is by induction on n. The case $n = 1$ follows easily from the fact that we have diagonalized over T_ξ's. So assume $n > 1$. We may assume that some fixed basic open set in X^n separates all elements of K because freeness is an open relation. We are identifying here an element of K with the corresponding n-tuple which is increasing with respect to the enumeration of Y. As before K is identified with the function g from a subset of X^{n-1} into X mapping the first $n - 1$ many elements of an s in K into its last element. Note that by the inductive hypothesis the family K_0 of all s in K for which the last element of s is in $\omega_g^*(s \mid n - 1)$ contains almost all members of K, i.e., $K \setminus K_0$ is of size $< 2^{\aleph_0}$. Let g_0 be a countable dense subfunction of g. Then $g_0 = f_\xi$ for some ξ so we can pick s in K_0 with all indexes above ξ and all the indexes of members of g_0. Let $\langle y, F(y) \rangle$ be the last element of s. Then

$$\langle y, F(y) \rangle \in \omega_{f_\xi}^*(s \mid n - 1)$$

implies that

$$F \cap \omega_{g_0}^*(s \mid n - 1)$$

is not connected, so we can pick two free members p and q of this set. Pick open I and J containing p and q respectively such that every element of I is free of every element of J. By the definition of $\omega_{g_0}^*$ $(s \mid n - 1)$ there is u in $\mathrm{dom}(g_0)$ free from $s \mid (n - 1)$ such that $g_0(u)$ is in I. Since $u \cup s \mid (n - 1)$ is free pick open U containing $s \mid (n - 1)$ such that every element of $\mathrm{dom}(g_0)$ in U is free from u. Pick a v in $U \cap \mathrm{dom}(g_0)$ such that $g_0(v)$ is in J. Then $u \cup \{g_0(u)\}$ and $v \cup \{g_0(v)\}$ are two members of K whose union is free. This finishes the proof.

The standard example of a space X for which $c(X)$ is inaccessible and non-attained ([11]) is obtained as a product of small spaces. It is well-known (see [84]) that any space obtained in this way has the inaccessible cardinal as a precaliber. In the next result we use the above diagonalization argument to obtain a quite different example of this kind.

4.5. THEOREM. *Let* $\theta = cf(2^{\aleph_0})$. *Then there is a compact zero-dimensional space* X *such that* X *has a disjoint family of open sets of any size* $< \theta$ *but none of size equal to* θ *and, moreover,* X^2 *has a disjoint family of open sets of size* θ.

PROOF. We shall closely follow the proof and notation of [58] so we assume the reader has a copy of [58] at hand. The translation $x \mapsto t + x$ of **R** will be denoted by t. For $A \subseteq \mathbf{R}$ set

$$t + A = \{t + x : x \in A\}.$$

Let $\{f_\xi : \xi < 2^{\aleph_0}\}$ enumerate all continuous mappings from G_δ subsets of $\mathbf{R}^n$, $(n < \omega)$ into **R**. Now, recursively on α define disjoint $1 - 1$ sequences

$\{r_\alpha : \alpha < 2^{\aleph_0}\}$ and $\{t_\alpha : \alpha < 2^{\aleph_0}\}$ of reals as follows. Fix $\alpha < 2^{\aleph_0}$, and let A_α be the closure of

$$\{r_\xi : \xi < \alpha\} \cup \{t_\xi : \xi < \alpha\}$$

under the functions from $\{f_\xi : \xi < \alpha\}$ and translations from $\{t_\xi : \xi < \alpha\}$. Pick t_α in $\mathbf{R} \setminus A_\alpha$ and r_α outside the closure of $A_\alpha \cup (t_\alpha + A_\alpha)$ under the functions from

$$\{f_\xi : \xi < \alpha\} \cup \{t_\xi : \xi \le \alpha\}.$$

It should be clear that [58] (or the proof of 4.2) shows that

$$\{r_\alpha : \alpha < 2^{\aleph_0}\} \cup \{t_\alpha : \alpha < 2^{\aleph_0}\}$$

is an entangled set in the following sense: For any disjoint family K of size continuum of finite subset of this set, all of some fixed size n, and for every $s \in 2^n$ there exist distinct F and G in K such that for all $i < n$, if x is the ith member of F and y is the ith member of G, then $x < y$ holds if and only if $s(i) = 1$. (We are considering here F and G enumerated according to the usual ordering of the real numbers.) Let C be a closed unbounded subset of $2^{\aleph_0}$ of type θ such that

$$\forall \, \gamma < \delta \text{ in } C \ \gamma + \gamma < \delta.$$

For $\gamma \in C$ let $\gamma(+)$ be the first member of C above γ. Put

$$E = \{\langle r_\alpha, t_{\gamma(+)} + r_\alpha\rangle : \gamma \in C \ \& \ \alpha \in [\gamma, \gamma + tp(C \cap \gamma))\}$$

considered as a subposet of $\mathbf{R}^2$ under coordinatewise ordering. Let X_0 be the set of all chains of E and let X_1 be the set of all antichains of E considered as subspaces of $\{0, 1\}^E$ under the standard identification. Finally, let

$$X = X_0 \oplus X_1.$$

Note that X_1 has a disjoint family of open sets of any size $< \theta$. Clearly, $X_0 \times X_1$ contains θ pairwise disjoint open sets (see [58]). So all that is left to be proved is that the posets P_0 and P_1 of finite chains and antichains of E, respectively are θ cc. The proof for P_0 is exactly the same as in [58], so we prove the fact only for the poset P_1.

So, let $\{p_\xi : \xi < \theta\}$ be a sequence of members of P_1. We may assume that p_ξ's are disjoint and of the same size $n \ge 1$ and that a fixed set of $2n$ rational intervals separates every p_ξ. Note that for a given γ in C, a p_ξ can have at most one pair of the form $\langle r_\alpha, t_{\gamma(+)} + r_\alpha\rangle$ for some α in

$$[\gamma, \gamma + tp(C \cap \gamma)).$$

So it suffices to find two $\xi \ne \zeta < \theta$ such that if the pair $\langle r_\alpha, t_{\gamma(+)} + r_\alpha\rangle$ in p_ξ corresponds to the pair $\langle r_\beta, t_{\delta(+)} + r_\beta\rangle$ in p_ζ, then

$$r_\alpha < r_\beta \text{ and } t_{\gamma(+)} - t_{\delta(+)} > r_\beta - r_\alpha.$$

This follows from the following general

CLAIM. *Suppose* $\mathcal{F} \subseteq [\{r_\alpha, t_\alpha : \alpha < 2^{\aleph_0}\}]^{2n}$, $(1 \leq n < \omega)$ *is a cofinal family such that any* $F \in \mathcal{F}$ *has the form*

$$\{r_{\alpha_0}, t_{\gamma_0}, \ldots, r_{\alpha_{n-1}}, t_{\gamma_{n-1}}\}$$

where $\alpha_0 < \gamma_0 < \cdots < \alpha_{n-1} < \gamma_{n-1}$. *Then for any* $\sigma, \tau \in \{0, 1\}^n$ *we can find two distinct*

$$F = \{r_{\alpha_i}, t_{\gamma_i} : i < n\} \text{ and } G = \{r_{\beta_i}, t_{\delta_i} : i < n\}$$

in $\mathcal{F}$ *such that for all* $i < n$:

(a) $\sigma_i = 0$ *iff* $r_{\alpha_i} < r_{\beta_i}$,
(b) $\tau_i = 0$ *iff* $t_{\gamma_i} < t_{\delta_i}$,
(c) $|r_{\alpha_i} - r_{\beta_i}| < |t_{\gamma_i} - t_{\delta_i}|$.

PROOF. Suppose the claim is false and let n (≥ 1) be the minimal integer for which it fails. Let $\mathcal{F}$ be the corresponding family. We shall identify an $F = \{r_{\alpha_i}, t_{\gamma_i} : i < n\}$ with

$$\langle r_{\alpha_0}, t_{\gamma_0}, \ldots, r_{\alpha_{n-1}}, t_{\gamma_{n-1}} \rangle \in \mathbf{R}^{2n}.$$

Define an f from a subset of $\mathbf{R}^{2n-1}$ into $\mathbf{R}$ by letting $\langle z, r \rangle \in f$ iff there is an $F = \{r_{\alpha_i}, t_{\gamma_i} : i < n\}$ in $\mathcal{F}$ such that

$$\langle z, r \rangle = \langle \langle r_{\alpha_0}, t_{\gamma_0}, \ldots, r_{\alpha_{n-1}} \rangle, t_{\gamma_{n-1}} \rangle.$$

The following fact gives the required contradiction since the sequence of t_α's diagonalizes all continuous functions from G_δ subsets of $\mathbf{R}^{2n-1}$ into $\mathbf{R}$.

Subclaim. $B_0 = \{z \in dom(f) : |\omega_f(z)| > 1\}$ *is not cofinal in* $2^{\aleph_0}$.

PROOF. Otherwise, by going to a cofinal subset of B_0 and by symmetry, we may assume to have $c < d$ in $\mathbf{Q}$ such that
(d) $\forall v \in B_0 \forall \epsilon > 0 \exists z \in dom(f)[\| v - z \| < \epsilon \ \& \ f(v) < c < d < f(z)]$.
For definiteness assume $\tau_{n-1} = 0$. Let $\epsilon_0 = d - c$. By the minimality of n and the proof of Theorem 1 of [58] (or the proof of 4.2) we can find u, v in B_0 which correspond, respectively, to

$$F = \{r_{\alpha_i}, t_{\gamma_i} : i < n\} \text{ and } G = \{r_{\beta_i}, t_{\delta_i} : i < n\}$$

from $\mathcal{F}$ as in the definition of f such that (a), (b) and (c) hold for all $i < n-1$, and

$$\sigma_{n-1} = 0 \text{ iff } r_{\alpha_{n-1}} < r_{\beta_{n-1}}.$$

Moreover, we may assume that

$$|u_i - v_i| < \epsilon_0 \text{ for all } i < 2n - 1$$

Let U be the set of all

$$z = \langle r_0, t_0, \ldots, r_{n-2}, t_{n-2}, r_{n-1} \rangle$$

in $\mathbf{R}^{2n-1}$ such that $\| u - z \| < \epsilon_0$ and for all $i < n - 1$:
(e) $\sigma_i = 0$ iff $r_{\alpha_i} < r_i$.

(f) $\tau_i = 0$ iff $t_{\gamma_i} < t_i$.

(g) $|r_{\alpha_i} - r_i| < |t_{\gamma_i} - t_i|$.

(h) $\sigma_{n-1} = 0$ iff $r_{\alpha_{n-1}} < r_{n-1}$.

Then U is an open set containing v. Since $\mathcal{F}$ contradicts the claim, it follows that for all $z \in \mathrm{dom}(f) \cap U$ either $f(z) < f(u) < c$, or else

$$|f(z) - f(u)| \leq |r_{\alpha_{n-1}} - r_{n-1}| < \epsilon_0.$$

So, in any case $f(z) < d$. But this clearly contradicts (d) for ϵ small enough so that the ϵ-ball around v is included in U. This completes the proof.

4.6. Remarks.

The diagonalization results of 4.2 and 4.4 will be used in §8 where a detailed historical remark will be given.

The result of 4.5 appeared in our note [62j] where the diagonalization argument was made on any linearly ordered set L having a dense set D such that $|L| = |2^D|$.

The first paper that deals with the $sup = max$ problem for a cardinal function seems to be that of Kurepa [29]. To describe one such result from [29], suppose T is a family of sets which is a tree under $\supset$ and which has the property that incomparable elements of T are disjoint sets. Let T^d be the family

$$\{X \setminus Y : X, Y \in T \text{ and } X \supset Y\}$$

and let

$$b'(T) = sup\{|D| : D \text{ is a disjoint subfamily of } T^d\}$$

The result of [29; p. 113] is that $b'(T)$ is attained if it is a singular cardinal. By looking at a partition tree of a linearly ordered space L (see [29] or [56]), an immediate consequence of this result is that a singular cellularity of a linearly ordered space is always attained. This result has been later extended to all spaces by Erdös-Tarski [11].

Finally, we note another application of the idea of 4.1. From the proof of 4.1, it should be clear that we can find a 1-1 mapping f from a subset of E_0 of size continuum into E_1 such that f contains an increasing subfunction of any size less than the continuum but f contains no monotonic subfunction of size equal to the continuum. Let Z be the graph of f, i.e., the set of all pairs of the form $\langle x, f(x) \rangle$. The usual euclidean topology of Z is refined by letting a neighborhood of an $\langle x, f(x) \rangle$ in Z be the intersection of a euclidean neighborhood and the set

$$\{\langle y, f(y) \rangle : y \leq x \,\&\, f(y) \geq f(x) \text{ or } y \geq x \,\&\, f(y) \leq f(x)\}.$$

An increasing subfunction of f is a discrete subspace of Z, so Z has a discrete subspace of any size less than the continuum. A subspace of Z weakly separated (see §3) by a sequence of neighborhoods with the same euclidean part is an increasing subfunction of f. Thus, Z has no weakly separated subspace of size continuum. This shows that if the continuum is a limit cardinal, then first countable zero-dimensional space Z witnesses sup $\neq$ max for several cardinal functions, simultaneously. Using a more involved diagonalization, a compact example with these properties can also be found.

5. (S) AND (L) AND THE SOUSLIN HYPOTHESIS

The Souslin Hypothesis is the same as the Conjecture (L) restricted to linearly ordered spaces. This has been shown by D. Kurepa (in the 1930's) by proving that any ccc linearly ordered space satisfies the hypothesis of (L) (see [52] for a set of references concerning this result). A more known result of Kurepa from the same period is that SH is the same as saying that every uncountable tree has an uncountable chain or antichain. The comparability relation of a tree is very close to the intersection relation on a family of intervals (or rather convex sets) of a linearly ordered set. This has been explained by the following two results from [53].

5.0. THEOREM. *The Souslin Hypothesis holds iff every family of intervals of a linearly ordered set with no uncountable disjoint subfamily is the union of countably many pairwise (and therefore finitely) intersecting subfamilies.*

5.1. THEOREM. (MA). *Every family of size less than continuum of intervals of a linearly ordered set without uncountable pairwise intersecting subfamily is the union of countably many disjoint subfamilies.*

In [55] we have shown that the form 5.0 of SH can be extended to arbitrary partially ordered sets in the following way.

5.2. **THEOREM.** (PFA). *Every partially ordered set P with no uncountable antichain is the union of countably many up-directed subsets P_n such that every countable subset of a P_n has an upper bound in P.*

No satisfactory extension of 5.1 is known, though if P is up-directed, the correct extension of 5.1 is an immediate consequence of Theorem 3 of [54].

The comparability relation on a Souslin tree T gives a particularly nice partition of $[T]^2$ into two cells. We shall now consider one more partition naturally determined by T because of its relevance to the Conjectures (S) and (L). In doing this we shall find it convenient to identify T with a partition

$$\tau : [\omega_1]^2 \to \omega$$

in such a way that T is equal to

$$\{\tau(\cdot, \beta) \mid \alpha : \alpha \leq \beta < \omega_1\}$$

with the inclusion ordering. We are using the convention that $[\omega_1]^2$ is the set of all *ordered* pairs $\langle \alpha, \beta \rangle$ with $\alpha < \beta < \omega_1$. As demonstrated in [60], this way of looking at trees can be very useful indeed. We shall also assume that every point t of our tree T has $\aleph_1$ successors and that $t^\wedge n$ is in T for all $n < \omega$.

5.3. LEMMA. *For every finite n and a disjoint family $\{F_\xi : \xi < \omega_1\}$ of n element subsets of ω_1 there is a $\beta < \omega_1$ such that for every $\gamma \geq \beta$ and $h : n \to \omega$ there is a $\xi < \beta$ such that if α is in F_ξ having ith place, then $\tau(\alpha, \gamma) = h(i)$.*

PROOF. Pick a countable elementary submodel M of $\mathbf{H}_{\aleph_2}$ containing everything relevant. We claim that $\beta = M \cap \omega_1$ works. To show this notice that for every $\gamma \geq \beta$, $\tau(\cdot, \gamma) \mid \beta$ is an M-generic branch of $T \cap M$, so all we need to check is an appropriate density fact. For this note that for any s in T, $h : n \to \omega$ and any F_ξ with minimum above the height of s we can find an extension t of s on level larger than $\max F_\xi$ such that if $\alpha \in F_\xi$ is in the ith place, then $t(\alpha) = h(i)$. This is an immediate consequence of the fact that $u^\wedge n$ is in T for all u in T and $n < \omega$. This completes the proof.

For $\alpha < \omega_1$ define $f_\alpha : \omega_1 \to \omega$ as follows. Let $f_\alpha(\alpha) = 1$; $f_\alpha(\xi) = 0$ for $\xi > \alpha$; $f_\alpha(\xi) = \tau(\xi, \alpha)$ for $\xi < \alpha$. Then 5.3 is saying that $\{f_\alpha : \alpha < \omega_1\}$ as a subspace of the Tychonoff product of ω_1 copies of ω is a very strong counterexample to (L). To state what we have just proved, let ST denote the fact that there is a Souslin tree, i.e. the negation of SH.

5.4. THEOREM. (ST). *There is a left-separated subspace X of $\{0, 1\}^{\omega_1}$ of type ω_1 such that every uncountable family of basic open subsets of $\{0, 1\}^{\omega_1}$ with disjoint domains contains a countable subfamily which covers an end-section of X.*

Subspace of $\{0, 1\}^{\omega_1}$ with this property are in the literature usually called weak-HFC's (see [21]).

We shall now see that the partition τ also hides a counterexample to (S). Hence, the Conjecture (S) also implies SH. This has been first shown by M. E. Rudin [29] using a construction which by an opinion of experts was "extremely complicated" [21] and for this reason remained isolated from the rest of the subject. So it seems appropriate giving here a simple construction showing this fact.

Fix a 1-1 sequence $\{r_\alpha : \alpha < \omega_1\}$ of elements of $\{0, 1\}^\omega$ and an enumeration $\{h_i : i < \omega\}$ of all maps

$$h : \{0, 1\}^n \times \{0, 1\}^n \to \omega,$$

where n is finite. Let n_i denote the integer corresponding to h_i and define

$$\phi : [\omega_1]^2 \to \omega$$

by

$$\phi(\alpha, \beta) = h_{\tau(\alpha,\beta)}(r_\alpha \mid n_{\tau(\alpha,\beta)}, r_\beta \mid n_{\tau(\alpha,\beta)}).$$

5.5. LEMMA. *For every uncountable $A \subseteq \omega_1$ there is a $\beta < \omega_1$ such that for every finite $F \subseteq \omega_1 \setminus \beta$ such that $\tau(\cdot, \gamma)$'s for γ in F agree below β and for every $g : F \to \omega$ there is an α in $A \cap \beta$ such that $\phi(\alpha, \gamma) = g(\gamma)$ for all γ in F.*

PROOF. Again we fix a countable elementary substructure M of $\mathbf{H}_{\aleph_2}$ containing everything relevant and show that $\beta = M \cap \omega_1$ works. So let F and g be as above and choose a finite n such that $r_\gamma \mid n$ ($\gamma \in F$) are distinct. Let $x \in \{0, 1\}^n$ be such that the set A_0 of all α in A for which r_α extends x is uncountable. Pick an h mapping $\{0, 1\}^n \times \{0, 1\}^n$ into ω such that

$$h(x, r_\gamma \mid n) = g(\gamma) \text{ for all } \gamma \text{ in } F.$$

Let i be such that $h = h_i$ and let $t \in T_\beta$ be such that $t \subseteq \tau(\cdot, \gamma)$ for all γ in F. Then since t is an M-generic branch of $T \cap M$ we can find an α in $A_0 \cap M$ such that $t(\alpha) = i$. It is clear now that $\phi(\alpha, \gamma) = g(\gamma)$ for all γ in F. This finishes the proof.

For $\alpha < \omega_1$ define $g_\alpha : \omega_1 \to 2$ as follows. First we put $g_\alpha(\alpha) = 1$. Then we put $g_\alpha(\xi) = 0$ if $\xi < \alpha$ or if $\tau(\cdot, \xi)$ and $\tau(\cdot, \alpha)$ are incomparable. In all other cases, we let

$$g_\alpha(\xi) = \min\{1, \phi(\alpha, \xi)\}.$$

Then $Y = \{g_\alpha : \alpha < \omega_1\}$ as a subspace of $\{0, 1\}^{\omega_1}$ is a counterexample to (S). To show this it is enough to prove that it does not contain uncountable discrete subspace. So suppose we are given increasing sequence $\{\alpha_\xi : \xi < \omega_1\}$ of ordinals from ω_1 and a sequence $\{h_\xi : \xi < \omega_1\}$ of finite partial functions from ω_1 into 2 such that g_{α_ξ} is in the basic open set $[h_\xi]$. Let $F_\xi = \mathrm{dom}h_\xi$. By refining the sequences we may assume that for every ξ, $\tau(\cdot, \gamma)$'s for γ in F_ξ agree below ξ. Letting $A = \{\alpha_\xi : \xi < \omega_1\}$ and applying the proof of 5.5 we get the existence of $\xi < \eta$ such that $g_\xi \in [h_\eta]$. Thus, we have proved the following.

5.6. THEOREM. (ST). *The Conjecture (S) implies the Souslin Hypothesis.*

5.7. Remarks.

Some historical remarks concerning Kurepa's early results on SH, (L) and (S) can be found in [30], [31] and [56] besides the papers mentioned in this section.

The result of Rudin was also presented in [40; Ch. V]. We suggest the reader examine normality and other covering properties of the space Y of 5.6. Note that if the tree T is a full Souslin tree, in the terminology of [56;

§6], then the space Y is hereditarily separable in all of its finite powers and, in fact, it is a weak HFD in the terminology of [21]. In this case the partition ϕ has the following strong property: For every uncountable disjoint family $\mathscr{F}$ of finite subsets of ω_1, all of the same size n, and for every mapping h from $n \times n$ into ω there exist F and G in $\mathscr{F}$ such that if α is in the ith place in F and if β is in the jth place in G then $\phi(\alpha, \beta)$ is equal to $h(i, j)$.

6. (S) AND (L) AND LUZIN SPACES

Let I be an ideal on a set X. Then a subset Y of X is I Luzin if Y is uncountable but $Y \cap N$ is countable for all N in I. If X is a topological space and if I is the ideal of meager subsets of X, then an I Luzin set is simply called a *Luzin subset of X*. If X is a measure space and if I is the ideal of measure zero sets, then I Luzin sets are called *Sierpinski subsets of X*. In this section we shall be primarily interested in Luzin subsets of the space $\{0, 1\}^A$ with the usual topology and product measure.

The existence of I Luzin sets for most of ideas I is in fact a quite unlikely (though consistent) event. There are many reasons for this, but the main one seems to be the requirement that $Y \cap N$ is *countable* for all N in I rather than that it has cardinality less than some other fixed cardinal. For example, if I is the ideal σ-generated by compact subsets of the Baire space ω^ω than any unbounded subset A of ω^ω which is well-ordered by $<^*$ having a regular order type has the property that $N \cap A$ is of smaller size than A for all N in I.

Our interest here in Luzin subsets of $\{0, 1\}^A$ comes from the fact that they give very canonical counterexamples to the conjecture (L) when A is uncountable. To explain this let I_0 be the ideal on $\{0, 1\}^A$ σ-generated by the sets of the form

$$\{f \in \{0, 1\}^A : \forall n < \omega \; p_n \not\subset f\},$$

where p_n is a sequence of partial functions from A into 2 with disjoint domains all of the same finite size. An I_0 Luzin set is in the literature usually called an HFC-set [21]. It should be clear that any I_0 Luzin set is hereditarily Lindelöf in a strong sense and that by an inessential change it can be made nonseparable. It should also be clear that I_0 is a subset of both meager and measure zero ideals of $\{0, 1\}^A$. Thus every Luzin or Sierpinski set in $\{0, 1\}^A$ for some uncountable A gives a strong counterexample to (L). The existences of Luzin and of Sierpinski sets of reals have been deduced from the Continuum Hypothesis a long time ago (see [27]) using extremely simple diagonalization arguments. Let us give an illustration of such an argument by proving the following:

6.0. PROPOSITION. (MA). *There is a sequence $L = \{r_\alpha : \alpha < 2^{\aleph_0}\}$ of reals such that for every first category set M in some finite power $\mathbf{R}^k$ of the reals there is a $\beta < 2^{\aleph_0}$ such that $L \cap M$ contains no k-tuple with all indexes above β.*

PROOF. Let $\{M_\xi : \xi < 2^{\aleph_0}\}$ enumerate all F_σ first category subsets of $\mathbf{R}^k$, where $1 \le k < \omega$. The αth element r_α of L is chosen such that

(a) r_α is not in any first category set of the form $(M_\xi)_p$, where $\xi < \alpha$ and where p is a finite sequence of elements of $\{r_\xi : \xi < \alpha\}$.

(Here, $(M_\xi)_p$ denotes the set of all reals r such that $p \, {}^\wedge r$ is in M_ξ.)

Suppose there is a k and a F_σ first category set $M \subseteq \mathbf{R}^k$ such that $L \cap M$ contains cofinally many k-tuples. Let k be minimal with this property and fix a $\bar\xi$ such that $M = M_{\bar\xi}$. Pick also a subset S of $L \cap M$ such that for any $\alpha < 2^{\aleph_0}$ there is an s in S with all indexes of its members above α. By taking a transformation of $\mathbf{R}^k$, we may assume that the members of S are index-increasing. By the category version of the Fubini Theorem (see [27] and [105]), the set

$$C = \{p \in \mathbf{R}^{k-1} : (M_{\bar\xi})_p \text{ is first category}\}$$

is comeager. By the minimality of k, there is an s in S such that

$$p = s \mid (k-1)$$

is in C and such that all the indexes of members of s are above $\bar\xi$. But this contradicts (a) with α equal to the index of $s(k-1)$. This completes the proof.

By changing the category with measure in the above proof, we get the following.

6.1. PROPOSITION. (MA). *There is a sequence $S = \{s_\alpha : \alpha < 2^{\aleph_0}\}$ of reals such that for every measure zero subset N of some $\mathbf{R}^k$ there is a $\beta < 2^{\aleph_0}$ such that $S \cap N$ contains no k-tuple with all indexes above β.*

6.2. THEOREM. *There is a Luzin set of reals iff there is a Luzin subset of $\{0, 1\}^{\omega_1}$.*

PROOF. The converse implication is clearly trivial, so we prove only the direct implication.

So fix a Luzin set $L = \{r_\alpha : \alpha < \omega_1\}$ in $\{0, 1\}^\omega$. As in §2 we fix a sequence $e : [\omega_1]^2 \to \omega$ of enumerations such that for all $\alpha < \omega_1$:

(a) $e_\alpha = e(\cdot, \alpha)$ is a 1–1 function from α into ω,
(b) $\{e_\beta \mid \alpha : \alpha \le \beta < \omega_1\}$ is countable.

For $\alpha < \omega_1$, set $f_\alpha(\xi) = r_\alpha(e(\xi, \alpha))$ if $\xi < \alpha$; otherwise $f_\alpha(\xi) = 0$. We claim that $\{f_\alpha : \alpha < \omega_1\}$ is Luzin in $\{0, 1\}^{\omega_1}$.

So, fix a sequence $\{p_n : n < \omega\}$ of finite partial functions from ω_1 into 2 such that the union of $[p_n]$'s is dense open in $\{0,1\}^{\omega_1}$ but it misses an uncountable subset $\{f_\alpha : \alpha \in A\}$ of L. Pick countable elementary submodel M of $H_{\aleph_2}$ containing everything relevant and let $\delta = M \cap \omega_1$. Fix an uncountable subset B of A and a mapping $\bar{e}$ from δ into ω such that

(c) $\bar{e} \subset e_\beta$ for all $\beta \in B$.

For $n < \omega$ define $q_n : D_n \to 2$ by

(d) $D_n = \{\bar{e}(\xi) : \xi \in \mathrm{dom}(p_n)\}$, and

(e) $q_n(\bar{e}(\xi)) = p_n(\xi)$ for ξ in $\mathrm{dom}(p_n)$.

Let V be the union of $[q_n]$'s. Then it is easily checked that V is dense open in $\{0,1\}^\omega$. Since L is Luzin in $\{0,1\}^\omega$ there must be a β in B such that $r_\beta \in V$. So there is an n so that r_β extends q_n. But by the definition of f_β and (c), (d) and (e), this means that f_β extends p_n, a contradiction. This completes the proof.

It should be clear that the method of proof of 6.2 also gives the following

6.3. THEOREM. *There is a Sierpinski set of reals iff there is a Sierpinski subset of* $\{0,1\}^{\omega_1}$.

A set of reals L is *strongly Luzin* if L is uncountable but for every first category subset M of some $\mathbf{R}^k$, the set $M \cap L^k$ contains no uncountable family of disjoint k-tuples. (Two k-tuples p and q are *disjoint* if $p_i \neq q_j$ for all $i, j < k$). Similarly one defines the notion of strongly Sierpinski by replacing the category with measure. Note that the set L of 6.0 is strongly Luzin if CH holds, and that S of 6.1 is strongly Sierpinski under CH.

6.4. PROPOSITION. *Suppose there is a set of reals which is strongly Luzin. Then there is a* $c : [\omega_1]^2 \to \omega_1$ *such that for every infinite disjoint* $\mathcal{F} \subseteq [\omega_1]^n$ *and uncountably disjoint* $\mathcal{G} \subseteq [\omega_1]^k$ *there is an F in $\mathcal{F}$ such that for any $h : n \times k \to \omega_1$ there is a G in $\mathcal{G}$ such that $c(F_i, G_j) = h(i,j)$ for all $i < n$ and $j < k$.*

PROOF. (Here F_i and G_j denote the ith and jth element of F and G, respectively, in the natural order.) Fix a strongly Luzin set

$$L = \{r_\alpha : \alpha < \omega_1\}$$

in ω^ω and choose a sequence $e : [\omega_1]^2 \to \omega$ of enumerations as in the proof of 6.2. The partition c is defined by

$$c(\alpha, \beta) = e_\beta^{-1}(r_\beta(e(\alpha, \beta))),$$

when $r_\beta(e(\alpha, \beta))$ is in the range of e_β; otherwise $c(\alpha, \beta) = 0$. Assume c doesn't satisfy 6.4 and fix counterexamples $\mathcal{F}$ and $\mathcal{G}$. This means that for every F in $\mathcal{F}$ there is a

$$h_F : n \times k \to \omega_1$$

such that for all G in $\mathcal{G}$,

(1) $\exists\, i < n\; \exists\, j < k\;\; c(F_i, G_j) \neq h_F(i, j)$.

We may assume that for some δ, $\mathcal{F} \subseteq [\delta]^n$ and range $h_F \subseteq \delta$ for all F in $\mathcal{F}$. By shrinking $\mathcal{G}$ we may assume that there is a sequence e^j, $(j < k)$ of mappings from δ into ω such that for all G in $\mathcal{G}$ and $j < k$,

(2) if $\alpha = G_j$, then $e^j \subset e_\alpha$.

Set

$$M = \{p \in \mathbf{R}^k : \forall\, F \in \mathcal{F}\; \exists\, i < n\; \exists\, j < k\;\; p_j(e^j(F_i)) \neq e^j(h_F(i, j))\}.$$

Then M is a nowhere dense subset of $\mathbf{R}^k$ such that $M \cap L^k$ contains

$$\{\langle r_{G_j} : j < k\rangle : G \text{ in } \mathcal{G}\}$$

as an uncountable set of disjoint k-tuples. This contradicts the fact that L is strongly Luzin and finishes the proof.

Let c be as in 6.4 and define subspaces

$$X = \{f_\alpha : \alpha < \omega_1\} \text{ and } Y = \{g_\alpha : \alpha < \omega_1\}$$

of $\{0, 1\}^{\omega_1}$ as follows. First let $f_\alpha(\alpha) = 1$ and $g_\alpha(\alpha) = 1$. If $\xi < \alpha$, then $f_\alpha(\xi)$ is equal to 0 and $g_\alpha(\xi)$ is equal to the minimum of 1 and $c(\xi, \alpha)$. If $\xi > \alpha$, then $g_\alpha(\xi)$ is equal to 0 and $f_\alpha(\xi)$ is equal to the minimum of 1 and $c(\alpha, \xi)$. Then it is easily seen that X and Y satisfy the conclusions of the following two corollaries, respectively.

6.5. COROLLARY. *Suppose there is a set of reals which is strongly Luzin. Then there is a right-separated subspace X of $\{0, 1\}^{\omega_1}$ of type ω_1 such that every infinite subset of X intersects every basic open subset of $\{0, 1\}^{\omega_1}$ with domain above some fixed ordinal in ω_1.*

6.6. COROLLARY. *Suppose there is a set of reals which is strongly Luzin. Then there is a left-separated subspace Y of $\{0, 1\}^{\omega_1}$ of type ω_1 such that every infinite family of basic open subsets of $\{0, 1\}^{\omega_1}$ with disjoint domains covers an end-section of Y.*

Thus, the simple diagonalization argument of 6.0 (under CH) gives us rather strong counterexample to (S) and (L). A version of 6.4 can be proved also for the notion of a strongly Sierpinski set of reals. This result (which we leave to the reader to state and prove) will be strong enough to give spaces X and Y satisfying 6.5 and 6.6, respectively.

Recall the definition of the set mapping H from §2 defined on some set $A = \{r_\alpha : \alpha < \omega_1\}$ of elements of ω^ω:

$$H(r_\beta) = \{r_\alpha : \alpha < \beta \;\&\; e(\alpha, \beta) \leq r_\beta(\Delta(r_\alpha, r_\beta))\}.$$

Note that the proof of 2.0 also shows the following which is of interest to us in the present context.

6.7. LEMMA. *For every second category set* $B \subseteq A$ *there exist* a *and* b *in* B *such that* $a \in H(b)$.

Hence, if we start with a Luzin set A the associated set mapping H has no uncountable free sets. So we may use H to refine the topology of A as in §2. What we want to point out here is that the refinement of a Luzin set topology usually has some additional properties and allows more modifications and the reader is urged to examine this space carefully. For example, a modification of this space gives a normal space whose product with the unit interval is not normal.

A *Luzin* space is any uncountable *Hausdorff* space X without isolated points such that every nowhere dense subset of X is countable. Constructing a *regular* and *nonseparable* Luzin space has been a traditional way in refuting the Conjecture (L) using various principles such as CH. This follows from the obvious fact that any Luzin space is hereditarily Lindelöf. The following result shows that, in fact, (L) can be refuted assuming the existence of *any* Luzin space not necessarily regular and nonseparable.

6.8. THEOREM. *If (L) holds, then there are no Luzin spaces.*

PROOF. Let X be a Luzin space. We shall construct a 1–1 sequence $Y = \{x_\alpha : \alpha < \omega_1\}$ of elements of X and for each α an open set U_α containing x_α such that

(a) $x_\xi \notin U_\alpha$ for all $\xi < \alpha$,
(b) $Y \cap (\bar{U}_\alpha \setminus U_\alpha) = \phi$.

Then Y with the topology generated by

$$\{U_\alpha \cap Y, Y \setminus U_\alpha : \alpha < \omega_1\}$$

is a nonseparable *regular* hereditarily Lindelöf space being a continuous image of Y with the topology induced from X. Clearly, Y with the new topology can't be separable and is therefore a counterexample to (L). This will clearly finish the proof of 6.5.

To prepare for our construction of Y we first observe that the regular open algebra of X can't be a Souslin algebra since (L) implies the Souslin Hypothesis. So, we can find uncountable subset X_0 of X and continuous one-to-one $f : X_0 \to \mathbf{R}$. By removing all isolated points of X_0 one gets a Luzin space, therefore we may assume that, in fact, X_0 is equal to X.

CLAIM. *For every countable subset D of X there is an open set V containing D such that $X \setminus V$ is uncountable.*

PROOF. Otherwise, since f is continuous, every open set of reals containing $f''D$ contains all but countably many elements of $f''X$. Hence there is an uncountable concentrated set of reals. By an old result of Rothberger ([38], [8]) this means that ω^ω contains a subset of size $\aleph_1$ which is unbounded with respect to the ordering of eventual dominance. By the result of 0.6 this contradicts (L).

The sequences of x_α's and U_α's are constructed recursively on α as follows. Let

$$D_\alpha = \{x_\xi : \xi < \alpha\} \cup \bigcup \{\bar{U}_\xi \setminus U_\xi : \xi < \alpha\}.$$

Since X is Luzin, D_α is countable. By the Claim there is an open set V_α containing D_α such that $F_\alpha = X \setminus V_\alpha$ is uncountable. Since F_α can't be nowhere dense it contains a nonempty open set U_α. Let x_α be any point of U_α. It is obvious that the conditions (a) and (b) remain satisfied. This completes the proof.

Note the following corollary of the proof.

6.9. THEOREM. *If there is a hereditarily separable Luzin space, then there exist first countable counterexamples to (S) and (L).*

Concerning 6.8 it would be interesting to know whether (S) also implies the nonexistence of Luzin spaces. It might be the case that (S) and/or (L) imply Martin's axiom for σ-centered posets or even the full Martin's axiom at the first uncountable level.

6.10. Remarks.

The observation that any Luzin space is hereditarily Lindelöf seems to have been first made by Amirdzanov and Shapirovskii [97] and by White [66]. The paper [97] also gives an extension of the classical constructions of Luzin sets of reals (due to Mahlo [102] and Luzin [103]) to a large class of Baire spaces.

The numerous results of Hajnal and Juhász about the HFD's and HFC's are surveyed in [21] and [22]. (Note that the counterexamples to (S) and (L) in 6.5 and 6.6 are strong HFD's and HFC's in the terminology of [21].)

In [66] White noticed that a Sierpinski set of reals is a Luzin space in the density topology of the reals (see [105]). This gave the first CH-counterexample to (L), (see [50]). In [6], van Douwen, Tall and Weiss use some other *nonseparable regular* Luzin spaces to give counterexamples to (L).

In [26], Kunen proves that MA plus non-CH imply that there are no Luzin spaces.

The results of 6.0 to 6.4 first appeared in our note [62a]. The same stepping-up method is applicable to many other ideals on $\{0,1\}^\omega$ and to many other combinatorial objects on ω. In the literature one can find many results involving stepping-up combinatorial objects from ω to ω_1 (see [98; p. 461], [36], [99], ...), but the advantage of using special Aronszajn tree as a stepping-up tool seems to have been first realized in [100] and [62a].

The result of 6.8 first appeared in our note [62h]. Concerning its proof, we should note that it is a rare instance when one is able to extract a regular space from a nonregular one.

In view of the connections pointed out in this section, it is natural to expect that many examples of spaces relevant to (S) and (L) can be found in the forcing extension obtained by adding one Cohen or one random real c in $\{0, 1\}^\omega$. First examples of this kind have been constructed by Roitman and Kunen in [36]. As an illustration of one such construction, let us reformulate Theorem 4 of [59] as follows. For r in $\mathbf{V} \cap \{0, 1\}^\omega$, define r_c in $\{0, 1\}^\omega$ by

$$r_c(n) = r(n) + c(r|n) \ (mod \ 2),$$

where $c : \{0, 1\}^{<\omega} \to 2$ is a fixed Cohen or random real over $\mathbf{V}$. Then the proof of Theorem 4 of [59] shows that

$$S = \{r_c : r \text{ in } \mathbf{V} \cap \{0, 1\}^\omega\}$$

is an $\aleph_1$-entangled set of reals of size $2^{\aleph_0}$ in the following sense (see [58]): For every finite $n \geq 1$, for every uncountable set F of increasing n-tuples of elements of S, and for every ϵ in $\{0, 1\}^n$ there exist s and t in F such that

$$\forall \ i < n \ (s_i < t_i \Leftrightarrow \epsilon_i = 0).$$

The set S can be used in constructions of many examples relevant to problems we have been considering here. For example, using S we can construct (see [58; Theorem 6]) a compact ccc space X whose square contains $2^{\aleph_0}$ disjoint open sets. Another example, relevant to §3, is obtained by considering the Sorgenfrey topologies $S[\leq]$ and $S[\geq]$ on S (see §0; $\leq$ is the usual (lexicographical) ordering of the Cantor set). To see this, note that by the fact that S is $\aleph_1$-entangled, every finite power of $S[\leq]$ or $S[\geq]$ is both hereditarily Lindelöf and hereditarily separable. On the other hand, the product

$$S[\leq] \times S[\geq]$$

contains a closed discrete subspace of size $2^{\aleph_0}$ and therefore is not normal.

If $c : \omega^{<\omega} \to \omega$ is a Cohen real and if for r in $\mathbf{V} \cap \omega^\omega$ we define now r_c in ω^ω by $r_c(n) = c(r|c(r|n))$, then the set S has the following property: Every uncountable subset of S contains an arbitrarily large finite subset which splits at the same place. For each n fix a family I_i^n, $(i \leq 2^{n+2})$ of open intervals of lengths $\leq 2^{-n-1}$ which covers $[0, 1]$. For s in S let A_s be the set of all x in $[0, 1]$ such that x is not in $I_{s(n)}^n$ for almost all n in ω. Then each A_s has measure zero but every uncountable subfamily of $\{A_s\}$ covers $[0, 1]$. (Compare this with [108] and [109].) All this shows that, in $\mathbf{V}[c]$, Martin's axiom fails in a strong sense.

7. FORCING AXIOMS FOR *ccc* PARTITIONS

Let X be a regular space for which the conclusion of (S) fails. Then we can find an uncountable $Y \subseteq X$ with a well-ordering $<$ and for every x in Y a neighborhood U_x of x in X such that y is not in $\bar{U}_x$ for all $y > x$ in Y. In order to prove that (S) holds for X we have to show that X does not satisfy the hypothesis of (S). A natural way for showing this is to produce an uncountable $D \subseteq Y$ such that x is not in U_y for all $x < y$ in D. In other words we have to produce an uncountable zero-homogeneous set for the partition

(1) $$[Y]^2 = K_0 \cup K_1$$

determined by:

$$\{x, y\}_< \in K_0 \text{ iff } x \notin U_y.$$

In fact, (S) is equivalent to such a Ramsey-type statement. To see this we need to make a closer examination of this partition. First of all, we may assume $Y = X$ and that X as a set is equal to ω_1. Note that this means that X is a zero-dimensional space so we choose U_x's to be clopen. For x in X let U_x^1 be equal to U_x and U_x^0 be equal to the complement of U_x. Let $\hat{X}$ denote the weaker topological space generated by

$$\{U_x^0, U_x^1 : x \in X\}.$$

If X contradicts to (S), then $\hat{X}$ also contradicts (S), so we need to produce an uncountable discrete subspace of $\hat{X}$. By reformulating this fact in terms of the partition (1), we get the following reformulation of (S):

(S_p) For every partition $[\omega_1]^2 = K_0 \cup K_1$ there is an increasing sequence $\{a_\xi : \xi < \omega_1\}$ of n element subsets of ω_1 and a $k < n$ auch that for all $\xi < \zeta$, either there is an $i < k$ such that $\{a_{\xi k}, a_{\zeta i}\} \in K_1$ or for some $i \geq k$, $\{a_{\xi k}, a_{\zeta i}\} \in K_1$ iff $\{a_{\zeta k}, a_{\zeta i}\} \in K_0$.

(Here, n is finite, $a_{\xi i}$ is the ith element of a_ξ in the increasing enumeration of a_ξ, and we are using the convention that $\{\alpha, \alpha\} \in K_1$ for all α.) If we start with a sequence V_x, $(x \in X)$ related to (1) by the formula

$$\{x, y\}_< \in K_0 \text{ iff } y \notin V_x$$

and then reformulate the fact that the corresponding space has an uncountable discrete subspace, we obtain the following reformulation of (L):

(L_p) For every partition $[\omega_1]^2 = K_0 \cup K_1$ there is an increasing sequence $\{a_\xi : \xi < \omega_1\}$ of n element subsets of ω_1 and a $k < n$ such that for all $\xi < \eta$ either there is an $i > k$ such that $\{a_{\xi i}, a_{\eta k}\} \in K_1$ or else for some $i \leq k$, $\{a_{\xi i}, a_{\eta k}\} \in K_1$ iff $\{a_{\xi i}, a_{\xi k}\} \in K_0$.

This explains our previous claim that (S) and (L) are, in fact, two Ramsey-type properties of the uncountable. The purpose of this and the next section is to explain some methods in producing homogeneous sets for the partition (1) in various situations. This will be presented using the language of forcing axioms stated in the forms of Ramsey-type statements about the uncountable.

Suppose S is an uncountable set and that $[S]^{<\omega} = K_0 \cup K_1$ is a given partition. Then we shall say that this is a *ccc partition* if

(a) $\{x\}$ is in K_0 for every element x of S.
(b) A subset of an element of K_0 is also in K_0.
(c) Every uncountable subset of K_0 has two elements whose union is in K_0.

If m is finite, then a partition of the form $[S]^m = K_0 \cup K_1$ is a *ccc* partition iff

(d) Every uncountable family of finite 0-homogeneous sets contains two members whose union is also 0-homogeneous.

In this section we shall present several applications of the following two Ramsey properties of the uncountable:

($\mathcal{L}$) *Let S be a given set of size less than the continuum and let $[S]^{<\omega} = K_0 \cup K_1$ be a given ccc partition. Then S can be covered by countably many subsets S_n such that for all n, $[S_n]^{<\omega} \subseteq K_0$.*

($\mathcal{H}$) *For every uncountable set S and every ccc partition $[S]^{<\omega} = K_0 \cup K_1$ there exists uncountable subset H of S such that $[H]^{<\omega} \subseteq K_0$.*

The following two results of [63] show that ($\mathcal{L}$) and ($\mathcal{H}$) are nothing more nor less than the two very familiar forcing axioms.

7.0. THEOREM. *($\mathcal{L}$) is equivalent to Martin's Axiom.*

7.1. THEOREM. *($\mathcal{H}$) is equivalent to Martin's Axiom for $\aleph_1$ dense sets.*

These results make it reasonable to call the next statement a forcing axiom.

(RFA) *Suppose S is an uncountable set and $[S]^2 = K_0 \cup K_1$ is a given partition. If there is a forcing notion introducing an uncountable homogeneous set for this partition, then such a homogeneous set, in fact, exists.*

Whether RFA is a consistent statement is still an open problem. If in the statement of ($\mathcal{H}$) we consider only partitions of a fixed finite dimension m, then the new (weaker) statement will be denoted by ($\mathcal{K}_m$). Note that ($\mathcal{K}_2$), which will be denoted simply as (K), is closely related to the familiar statement that every *ccc* poset has the property K (see [12]). Similarly, we define ($\mathcal{L}_m$) for a fixed finite dimension m.

Let us first make several remarks concerning the proofs of 7.0 and 7.1 in [63]. One of the main difficulties in proving these two results was in associating a *ccc* partition of the form

(2) $$[\kappa]^{<\omega} = K_0 \cup K_1$$

to a *tower* $\{a_\xi : \xi < \kappa\}$ in $[\omega]^\omega$, $\subset^*$, i.e. a sequence of the property that $a_\xi \subset^* a_\zeta$ for all $\zeta < \xi < \kappa$. The partition (2) is defined in [63] by letting a finite subset F of κ to be a member of K_0 iff

$$\forall\, k < \omega\ |a_F \cap k| \geq |\Delta_F \cap k|,$$

where

$$a_F = \bigcap_{\xi \in F} a_\xi, \text{ and}$$

$$\Delta_F = \{min(a_\xi \Delta a_\zeta) : \xi \neq \zeta \text{ in } F\}.$$

It is fairly easy to check that (2) is indeed a *ccc* partition but the main point of defining (2) in this way is that we have the following fact.

7.2. CLAIM. $\bigcap_{\xi \in H} a_\xi$ *is infinite for every 0-homogeneous set H.*

Hence, the existence of a 0-homogeneous set which is cofinal in κ would extend the tower. Of course, this is a far less obvious way to extend a tower than the standard one (see [12]), but the following result (of [63]) shows that it was well worth the effort (see Problem 9 of [75; p. 91]).

7.3. THEOREM. *There is a ccc nonseparable compact space of size continuum.*

PROOF. Let t be the minimal cardinality of a nonextending tower. Fix such a tower $\{a_\xi : \xi < t\}$ and extend the notation to define a_F and Δ_F for all subsets of t. Identifying $\mathcal{P}(t)$ and $\{0,1\}^t$, set

$$X = \{F \subseteq t : \forall\, k < \omega\ |a_F \cap k| \geq |\Delta_F \cap k|\}.$$

Note that X is a closed subspace of $\{0,1\}^t$ and that by the Claim every member of X has size $< t$. Since $2^{<t}$ is equal to the continuum all that is left to be proved is the *ccc* property of X. The proof of this fact is almost identical to the proof of the *ccc* property of (2) (see [63]).

We believe that such a kind of space, associated with a cardinal where a part of Martin's axiom fails first, might be relevant to many other problems of general topology. This belief is based on the possibility of using the available instance of MA to *prove* (rather than build in) properties of the associated space.

Let us now give few more instances of such a line of applying MA. The first such construction is reproduced here from [59; §3] and it shows how to obtain a large continuous subfunction of a given function f mapping a set of reals S into the reals. It will be more convenient to work with the Baire space ω^ω instead of the reals. It will also be more convenient to assume f is one-to-one which for our purpose here is not a loss of generality. The associated partition

$$(3) \qquad\qquad [S]^3 = K_0 \cup K_1$$

is defined by letting an F of $[S]^3$ into K_0 iff for every a, b, c in F

$$\Delta(a, b) \neq \Delta(a, c)$$

implies

$$\Delta(f(a), f(b)) \neq \Delta(f(a), f(c)).$$

7.4. LEMMA. *(3) is a ccc partition.*

PROOF. Let $\mathcal{F}$ be an uncountable family of finite 0-homogeneous sets for the partition (3). To each H in $\mathcal{F}$ we associate $k_H < \omega$, $s_H, t_H \subseteq \omega^{k_H}$, and $g_H : s_H \to t_H$ such that:

   (a) $H \mid k_H = s_H$,
   (b) $(f''H) \mid k_H = t_H$,
   (c) $g_H(a \mid k_H) = f(a) \mid k_H$,
   (d) $\Delta(a, b) < k_H$ for all $a \neq b$ in H,
   (e) $\Delta(f(a), f(b)) < k_H$ for all $a \neq b$ in H.

Since there exist only countably many parameters, we can find two H_0 and H_1 in $\mathcal{F}$ with the same quadruple of parameters. It is now easily checked using (a)-(e) that $H_0 \cup H_1$ is 0-homogeneous. This proves the Lemma.

7.5. LEMMA. *If H is a 0-homogeneous subset of S, then $f \mid H$ is continuous.*

PROOF. If $f \mid H$ is not continuous, we would be able to find an a in H and a sequence $\{a_i\}$ of elements of H such that $\lim a_i = a$, but either $\lim f(a_i)$ doesn't exist or it is not equal to $f(a)$. In any case we could find $i < j$ such that

$$\Delta(a, a_i) < \Delta(a, a_j) \text{ and}$$

$$\Delta(f(a), f(a_i)) = \Delta(f(a), f(a_j))$$

contradicting $\{a_0, a_i, a_j\} \in K_0$.

7.6. THEOREM. $(\mathcal{K}_3)$. *Every function from an uncountable set of reals into the reals is continuous on an uncountable set.*

Note that if we have started with a 1–1 function from **R** into **R** that is not continuous on any set of size continuum (see [47], [27]), then the associated compact space of all 0-homogeneous sets satisfies Theorem 4.0. Let us also note that 3 is the smallest known dimension of a *ccc* partition whose homogeneous sets produce continuous subfunctions as in 7.6. This is also the case with the following result.

7.7. THEOREM. (K_3). 2^{ω_1}, $<_{lex}$ *is embeddable in the ultrapower* $\omega^\omega/\mathcal{U}$, $<_\mathcal{U}$ *for every nonprincipal ultrafilter* $\mathcal{U}$ *on* ω.

PROOF. Suppose we have two sequences $\{a_\alpha : \alpha < \omega_1\}$ and $\{b_\alpha : \alpha < \omega_1\}$ in ω^ω such that a_α's are $<^*$-increasing, b_α's are $<^*$-decreasing and

$$\forall\, \alpha\, \forall\, n\ a_\alpha(n) \le b_\alpha(n).$$

The partition

(4) $$[\omega_1]^3 = K_0 \cup K_1$$

is defined by: $\{\alpha, \beta, \gamma\} \in K_0$ iff for all n

$$\{[a_\alpha(n), b_\alpha(n)], [a_\beta(n), b_\beta(n)], [a_\gamma(n), b_\gamma(n)]\}$$

is not a disjoint family of intervals.

CLAIM. *(4) is a ccc partition.*

PROOF. Suppose $\{F_\xi : \xi < \omega_1\}$ is family of 0-homogeneous sets. By refining, we may assume all F_ξ's have the same size k and that for some m in ω, $\langle\langle s_i, t_i\rangle : i < k\rangle$ in $(\omega^m \times \omega^m)^k$, and all ξ, we have:

(a) If α is the ith element of F_ξ, then $s_i \subset a_\alpha$ and $t_i \subset b_\alpha$.
(b) If $\alpha < \beta$ are in F_ξ, then for all $n \ge m$, $a_\alpha(n) \le a_\beta(n) \le b_\beta(n) \le b_\alpha(n)$.

It is now clear that $F_\xi \cup F_\zeta$ is 0-homogeneous for all ξ and ζ.

Suppose $H \subseteq \omega_1$ is an uncountable 0-homogeneous set for (4), n is in ω, and consider the family of intervals

$$I_n = \{[a_\alpha(n), b_\alpha(n)] : \alpha \in H\}.$$

Let I_n be antilexicographically minimal interval in I_n and let

$$I_n^0 = \{I \in I_n : I \cap I_n \ne \varnothing\}, \text{ and}$$

$$I_n^1 = \{I \in I_n : I \cap I_n = \varnothing\}.$$

Then it is easily checked that I_n^i is a pairwise intersecting family of intervals for all $i < 2$. So, by Helly's theorem both

$$c(n) = min(\bigcap I_n^0), \text{ and}$$

$$d(n) = min(\bigcap I_n^1)$$

exist ($min\varnothing = 0$). Thus, an uncountable 0-homogeneous set for (4) produces us two functions c, d in ω^ω such that for every α there is an m such that for all $n \geq m$ either

$$a_\alpha(n) \leq c(n) \leq b_\alpha(n), \text{ or}$$

$$a_\alpha(n) \leq d(n) \leq b_\alpha(n).$$

This means that if $\mathcal{U}$ is any nonprincipal ultrafilter on ω one of the $[c]_\mathcal{U}$ or $[d]_\mathcal{U}$ separates $\{[a_\alpha]_\mathcal{U} : \alpha < \omega_1\}$ from $\{[b_\alpha]_\mathcal{U} : \alpha < \omega_1\}$ in the ultrapower $\omega^\omega / \mathcal{U}$.

Inductively on s in $2^{<\omega_1}$ construct a_s and b_s in ω^ω such that:

(c) $s \subset t$ implies $a_s <^* a_t <^* b_t <^* b_s$,

(d) $a_{s^\wedge 0} <^* b_{s^\wedge 0} <^* a_{s^\wedge 1} <^* b_{s^\wedge 1}$.

Fix a nonprincipal ultrafilter $\mathcal{U}$ on ω and define a mapping φ from 2^{ω_1} into $\omega^\omega / \mathcal{U}$ as follows: For a given f in 2^{ω_1}, by the first part of the proof, the sequences

$$\{[a_{f|\alpha}]_\mathcal{U} : \alpha < \omega_1\} \text{ and } \{[b_{f|\alpha}]_\mathcal{U} : \alpha < \omega_1\}$$

can be separated in $\omega^\omega / \mathcal{U}$, so let $\varphi(f)$ be any such separating point of $\omega^\omega / \mathcal{U}$. It is clear that so defined

$$\varphi : 2^{\omega_1}, <_{lex} \rightarrow \omega^\omega / \mathcal{U}, <_\mathcal{U}$$

is strictly increasing. This completes the proof.

7.8. THEOREM. (K). *Every subset of ω^ω of size $\aleph_1$ is bounded in ω^ω, $<^*$.*

PROOF. This result first appeared as Theorem 15 of [59], but for the convenience of the reader we sketch the argument. So fix an $F \subseteq \omega^\omega$ of size $\aleph_1$. We may assume $F = \{a_\xi : \xi < \omega_1\}$ where a_ξ's are $<^*$-increasing. Moreover, we may assume that each a_ξ is an increasing function. Define

(5) $$[\omega_1]^2 = K_0 \cup K_1$$

by

$$\{\alpha, \beta\}_< \in K_0 \text{ iff } \exists k \ a_\alpha(k) > a_\beta(k).$$

Then (5) is a *ccc* partition and this follows from the following general fact [59; §4].

CLAIM. *Suppose F is a closed set mapping defined on a set of reals X such that for some well-ordering $<_w$ of X and all x in X, $F(x)$ contains no point strictly above x in the ordering $<_w$. Then every uncountable family of finite F-free subsets of X contains two elements whose union is also F-free.*

PROOF. Let $\{p_\alpha : \alpha < \omega_1\}$ be a given family of finite free subsets of X. We may assume p_α's are disjoint and of the same size n. We shall identify p_α with the element of X^n which enumerates p_α according to the ordering $<_w$. By refining the sequence of p_α's, we may assume that a fixed sequence

$I_0, \ldots, I_{n-1}$ of rational intervals separates each p_α in the sense that for all $i \neq j < n$,

$$(p_\alpha)_i \in I_i \text{ and } F((p_\alpha)_j) \cap I_i = \varnothing.$$

Furthermore, we assume that

$$(p_\alpha)_i <_w (p_\beta)_i$$

for all $\alpha < \beta < \omega_1$ and $i < n$. Fix a $\gamma < \omega_1$ such that $\{p_\alpha : \alpha < \gamma\}$ is dense in $\{p_\alpha : \alpha < \omega_1\}$ in the topology induced from X^n. Fix also $\delta > \gamma$. By the hypothesis of the claim, for all $i < n$,

$$J_i = I_i \backslash F((p_\gamma)_i)$$

is an open set containing $(p_\delta)_i$. So there must be a $\beta < \gamma$ such that $(p_\beta)_i$ is an element of J_i for all $i < n$. Then it follows directly that $p_\beta \cup p_\gamma$ is F-free. This proves the claim.

If (5) has an uncountable 0-homogeneous set, then by 0.7, F must be bounded. This completes the proof.

7.9. THEOREM. ($\mathcal{L}_2$). *If there is a (κ, λ^*) gap in $\omega^\omega, <^*$ (or in $[\omega]^\omega, \subset^*$), then either one of the κ and λ is equal to the continuum, or else $\kappa = \lambda = \omega_1$.*

PROOF. Suppose there is a (κ, λ^*) gap $\langle A, B \rangle$ in $\omega^\omega, <^*$ such that neither of these two conditions are satisfied. Thus $\kappa, \lambda < 2^{\aleph_0}$ and (say) $\kappa \neq \omega_1$. By the previous result and Rothberger's characterization of (κ, ω^*) or (ω, λ^*) gaps (see [38] or [8]), we know that $\kappa \neq \omega$. (We are also assuming that κ and λ are regular.) Let $X = A \times B$ and define

(6) $$[X]^2 = K_0 \cup K_1$$

by

$$\{\langle a, b \rangle, \langle \bar{a}, \bar{b} \rangle\} \in K_0 \text{ iff } \max\{\Gamma(a, \bar{b}), \Gamma(\bar{a}, b)\} \leq \max\{\Gamma(a, b), \Gamma(\bar{a}, \bar{b})\}.$$

(See §3 for the definition of Γ.)

CLAIM. *(6) is a ccc partition.*

PROOF. Let $\{F_\xi : \xi < \omega_1\}$ be a sequence of 0-homogeneous sets for (6). Since $\kappa > \omega_1$, there is a c in A above any element appearing as a first coordinate of a member of some F_ξ. By refining we may assume that F_ξ's are disjoint and all of the same size k. Furthermore, we may assume that for some m, $\langle \langle s_i, t_i \rangle : i < k \rangle$ in $(\omega^m \times \omega^m)^k$, and all ξ, we have:

  (a) If $\langle a, b \rangle$ is the ith element of F_ξ (in the lexicographical enumeration of F_ξ), then $s_i \subset a$ and $t_i \subset b$.
  (b) If $\langle a, b \rangle$ is in F_ξ, then $\Gamma(a, b)$, $\Gamma(a, c)$ and $\Gamma(c, b)$ are all $\leq m$.

It is now easily checked that so refined sequence of the F_ξ's has the property that $F_\xi \cup F_\eta$ is 0-homogeneous for all ξ and η.

By $(\mathcal{L}_2)$, X is the union of countably many 0-homogeneous sets X_n, $(n < \omega)$. An argument similar to that from §3 shows that this splits the gap $\langle A, B \rangle$, a contradiction.

Let us now consider the influence of MA on (S) and (L). We shall later see that MA is far from solving (S) and (L), but some very special cases of (S) and (L) do indeed follow from $(\mathcal{H})$. Since the proofs of this kind of results are giving us some ideas about the difficulties in producing uncountable homogeneous sets for partitions associated with (S) and (L), we decide to give a short presentation of what MA can do about (S) and (L).

A sequence $\{x_\alpha : \alpha < \theta\}$ of elements of a space X is called *free* in X iff for all $\beta < \theta$

$$\overline{\{x_\alpha : \alpha \leq \beta\}} \cap \overline{\{x_\alpha : \beta < \alpha < \theta\}} = \varnothing$$

Clearly, every infinite free sequence in X contains a discrete subspace of the same size. Let us call X *countably tight* if for all $Y \subseteq X$ and x in $\bar{Y}$ there is a countable $Y_0 \subseteq Y$ such that x is in $\bar{Y}_0$. The basic result which connects these two notions, (see [90; p. 41] for the historical remark) is that a compact space is countably tight iff it contains no uncountable free sequence. One direction of this result is obvious. In fact, no countably tight Lindelöf space can contain an uncountable free sequence.

7.10 THEOREM. $(\mathcal{K})$. *Suppose X is a regular space which contains no uncountable free sequence in any of its finite powers. Then X contains no counterexample to (S) or (L).*

PROOF. We shall prove the result only for (S) since the case of (L) is symmetrical. So, let $Y, <$ be a right separated subspace of X of type ω_1. For each y in Y we fix neighborhoods $U(y)$ and $V(y)$ of y in X such that $\overline{U(y)} \subseteq V(y)$ and $V(y)$ contains no element of Y above y. As usual, define

(7) $[Y]^2 = K_0 \cup K_1$

by

$$\{x, y\}_< \in K_0 \text{ iff } x \notin U(y).$$

CLAIM. *(7) is a ccc partition.*

PROOF. Otherwise, fix a sequence $\{F_\alpha : \alpha < \omega_1\}$ of finite 0-homogeneous sets such that for no $\alpha \neq \beta$, $F_\alpha \cup F_\beta$ is 0-homogeneous. We may assume F_α's are strictly $<$-increasing and of the same size $n \geq 1$. For each α, let $\{x_\alpha^0, \ldots, x_\alpha^{n-1}\}$ be the increasing enumeration of F_α. Let $\langle F_\alpha \rangle$ denote the element $\langle x_\alpha^0, \ldots, x_\alpha^{n-1} \rangle$ of X^n. For $\beta < \omega_1$, set

$$W_\beta^0 = \{z \in X^n : \exists \, i, j < n \; z_i \in U(x_\beta^j)\},$$

$$W_\beta^1 = \{z \in X^n : \exists \, i, j < n \; z_i \in V(x_\beta^j)\}.$$

Then W_β^0 and W_β^1 are open subsets of X^n such that:

(a) $\forall \alpha \leq \beta \; \langle F_\alpha \rangle \in W_\beta^0$,
(b) $\forall \alpha > \beta \; \langle F_\alpha \rangle \notin W_\beta^1$.

Note that by our choice of $U(y)$'s and $V(y)$'s, the closure of W_β^0 is included in W_β^1. It follows that

$$\{\langle F_\alpha \rangle : \alpha < \omega_1\}$$

is a free sequence in X^n. This completes the proof of the claim and also the proof of 7.10.

7.11. COROLLARY. ($\mathcal{K}$). *The following conditions are equivalent for a regular space X:*

(1) $X^{\aleph_0}$ *contains no uncountable discrete subspace.*
(2) $X^{\aleph_0}$ *is hereditarily separable.*
(3) $X^{\aleph_0}$ *is hereditarily Lindelöf.*

7.12. COROLLARY. ($\mathcal{K}$). *A compact countably tight space contains no counterexample to (S) or (L).*

PROOF. Tightness is productive in the class of all compact spaces (see [81] or [71]).

The results 7.11 and 7.12 have been originally proved by Kunen [72] and Szentmiklóssy [48], respectively. It is interesting to note that the proof of [72] and our proof of 7.10 are technically almost identical, but that a large number of experts, well familiar with [72], have found 7.12 surprising when [48] appeared (see [21], [37], [41] and [75]). The most natural class of spaces satisfying the hypothesis of 7.10 are spaces which are countably tight and Lindelöf in all finite powers. Note also that another interpretation of the proof of 7.10 gives the following.

7.13. THEOREM. ($\mathcal{L}_2$). *Let X be a regular space which contains no uncountable free sequence in any of its finite powers. Suppose Y is a subset of X of size $< 2^{\aleph_0}$ and $\{\langle U_y, V_y \rangle : y \in Y\}$ is a sequence of pairs of open subsets of X such that U_y contains y and $\bar{U}_y \subseteq V_y$ for all y in Y. Suppose further that one of the following two conditions is satisfied:*

(a) $V_y \cap Y$ *is countable for all y in Y, or*
(b) $\{x \in Y : V_x \text{ contains } y\}$ *is countable for all y in Y.*

Then Y can be decomposed into countably many sets Y_n such that $x \notin U_y$ for all $x \neq y$ in Y_n.

Let us now give two applications of the partition property $(\mathcal{H})$ itself. First we would like to mention the following simple but basic result because it gives the best topological explanation of why Martin's axiom solves the Souslin Problem.

7.14. THEOREM. ($\mathcal{H}$). *Suppose X is a regular ccc nonseparable space. Then X contains either an uncountable free sequence or an infinite subset with no accumulation points.*

PROOF. Since X is not separable, we can construct an increasing sequence $\{D_\alpha : \alpha < \omega_1\}$ of countable subsets of X and a sequence $\{U_\alpha : \alpha < \omega_1\}$ of nonempty open subsets of X such that if D is the union of the D_α's, then

(a) $\bar{U}_\alpha \cap \bar{D}_\alpha = \varnothing$ for all α,
(b) if a finite intersection of U_α's is nonempty, then it contains an element of D.

By ($\mathcal{H}$), there is an uncountable subset A of ω_1 such that $\{U_\alpha : \alpha$ in $A\}$ has the finite intersection property. By reindexing D_α's, we may assume that for all β,

$$\{U_\alpha \cap D_{\beta+1} : \alpha \leq \beta, \ \alpha \text{ in } A\}$$

has the finite intersection property. If for some β in A,

$$\bar{D}_{\beta+1} \cap \bigcap \{\bar{U}_\alpha : \alpha \leq \beta, \ \alpha \text{ in } A\}$$

is empty, the second conclusion of 7.14 is easily derived. So we may assume that this intersection is nonempty for all β in A and we may therefore pick an element x_β from it. It is now easily checked that

$$\{x_\alpha : \alpha \text{ in } A\}$$

is an uncountable free sequence in X. This completes the proof.

Thus, under ($\mathcal{H}$) every countably tight compact ccc space is separable. We suggest the reader examine the example of 7.3 to see how much the assumption of countable tightness is needed for this result.

We finish this section by considering a problem in topological measure theory. To state our result, we need a standard definition from measure theory (see [42]). Let X be an *arbitrary* topological space and let $\langle X, \mu, \mathcal{M} \rangle$ be a measure space on X where $\mathcal{M} = Borel(X)$. Then μ is called a *regular Radon measure* on X if and only if:

(1) $\forall x \in X \ \exists$ *open* $U(x \in U$ & $\mu(U) < \infty)$,
(2) $\forall B \in \mathcal{M} \ \mu(B) = inf\{\mu(U) : U$ is open & $B \subseteq U\}$,
(3) $\forall B \in \mathcal{M} \ \mu(B) = sup\{\mu(K) : K \subseteq B$ & $K \in \mathcal{M}$ & K is compact$\}$,
(4) $\forall x \in X \ \mu(\{\bar{x}\}) = 0$.

7.15. THEOREM. ($\mathcal{H}$). *Every regular Radon measure is σ-finite.*

PROOF. Pick a maximal disjoint $\mathcal{K} \subseteq \mathcal{M}$ such that for every K in $\mathcal{K}$:

(a) K is compact in X,
(b) $\mu(U \cap K) > 0$ for any open U in X such that $U \cap K \neq \varnothing$.

It is easily seen that $\mathcal{K}$ has size at most $\aleph_1$ but we shall show that $\mathcal{K}$ is, in fact, countable. This will clearly finish the proof since every K in $\mathcal{K}$ has finite measure and since X minus $\bigcup \mathcal{K}$ has measure zero because it contains no compact Borel set of positive measure (see (3)).

Assume that $\mathcal{K}$ is not countable and let us work for a contradiction. By compactness and by (1), for any K in $\mathcal{K}$ we can fix an open set $U(K)$ of finite measure containing K.

Let $\mathcal{P}_0$ be the set of all finite $p \subseteq \mathcal{K}$ such that for all K in p the set

$$F_p(K) = K \setminus \bigcup \{U(L) : L \in p \ \& \ L \neq K\}$$

has positive measure.

CLAIM. $[\mathcal{K}]^{<\omega} = \mathcal{P}_0 \cup \mathcal{P}_1$ *is a ccc partition.*

PROOF. So let $\{p_\alpha : \alpha < \omega_1\} \subseteq \mathcal{P}_0$ be given. We may assume that p_α's form a Δ system with root, p and that

$$\left(\bigcup \{U(K) : K \in p_\alpha\}\right) \cap \left(\bigcup \{K : K \in p_\beta \setminus p\}\right) = \varnothing.$$

for all $\alpha < \beta < \omega_1$. This is so, because every open set of finite measure can intersect at most countably many members of $\mathcal{K}$ (see (b)). Since the restriction of the measure algebra on any K in p has the Knaster property, by going to a subsequence of the p_α's we may assume that for all K in p the family $\{F_{p_\alpha}(K) : \alpha < \omega_1\}$ has all pairwise intersections of positive measure. Thus, we may assume that, in fact, $p = \varnothing$. By a further shrinking, we may assume that for some fixed positive and finite ϵ and M, we have:

(c) $\mu(F_{p_\alpha}(K)) > \epsilon$ for all $\alpha < \omega_1$ and $K \in p_\alpha$.
(d) $\mu(\bigcup \{U(K) : K \in p_\alpha\}) < M$ for all $\alpha < \omega_1$.

Pick an integer n such that $n\epsilon \geq M$. For $\alpha < \omega_1$ define

$$I(\alpha) = \{\xi < \alpha : p_\xi \cup p_\alpha \notin \mathcal{P}_0\}.$$

We claim that $|I(\alpha)| < n$ for all α. Namely, for any ξ in $I(\alpha)$ there is an K_ξ in p_ξ such that

$$F_{p_\xi}(K_\xi) \subseteq_{ae} \bigcup \{U(K) : K \in p_\alpha\}$$

Hence

$$|I(\alpha)|\epsilon < \mu\left(\bigcup \{F_{p_\xi}(K_\xi) : \xi \in I(\alpha)\}\right) \leq \mu\left(\bigcup \{U(K) : K \in p_\alpha\}\right) < M,$$

which by the choice of n means that $n > |I(\alpha)|$.

By the Pressing Down Lemma pick stationary $A \subseteq \omega_1$ such that $\alpha \notin I(\beta)$ for all α and β in A. Then $p_\alpha \cup p_\beta \in \mathcal{P}_0$ for all α and β in A. This proves the Claim.

By ($\mathcal{H}$) pick uncountable $\mathcal{L} \subseteq \mathcal{K}$ such that $[\mathcal{L}]^{<\omega} \subseteq \mathcal{P}_0$. Then for each K in $\mathcal{L}$,

$$\mathcal{L}(K) = \{K \setminus U(L) : L \in \mathcal{L} \ \& \ L \neq K\}$$

is a family of closed subsets of K with finite intersection property. So, by the compactness of K we can fix an element $y(K)$ from $\bigcap \mathcal{L}(K)$. Let

$$Y = \overline{\{y(K) : K \in \mathcal{L}\}} \cap (\bigcup \{U(K) : K \in \mathcal{L}\})$$

Then Y is Borel. Note that for all K in $\mathcal{L}$

$$Y \cap U(K) = \overline{\{y(K)\}} \cap U(K)$$

which by (4) means that $\mu(Y \cap U(K)) = 0$. But any compact $L \subseteq Y$ is covered by finitely many open sets of the form $U(K)$ where K is in $\mathcal{L}$, therefore it must have measure zero. By (3), this means that $\mu(Y) = 0$. On the other hand any open set U containing Y must have infinite measure, since it intersects uncountably many members of K (see (b)). By (2), this means that $\mu(Y) = \infty$, a contradiction. This completes the proof.

7.16. Remarks.

The reformulation of (S) and (L) in terms of partition was made by Roitman [34].

The reader is referred to [63] for more information concerning the Ramsey theoretic aspects of Martin's axiom.

The RFA was first considered in [62b]. Concerning the dimension 2 in the statement of RFA, let us note that RFA^3 (in the obvious notation) is false (see [63]).

The conclusion of 7.7 was first deduced by Woodin [68] but only from the full MA$_{\aleph_1}$ rather than (K_3). Note that 7.7 shows that (K_3) implies $2^{\aleph_0} = 2^{\aleph_1}$ (a result first proved in [63] using different methods) and it is unknown whether (K) also implies this (see [12] and [63]). Let us note that 7.8 is the only "combinatorial" (see [73]) consequence of MA$_{\aleph_1}$ we know of which follows from (K). To get the feeling of the difficulties involved, we suggest the reader try to specialize an Aronszajn tree using only (K_2). Again, it was known (see [63]) that (K_3) suffices for this task and only recently the author has been able to show that indeed (K_2) implies that all Aronszajn trees are special.

Concerning the Claim from the proof of 7.8, we should note that the proof of Theorem 1.4 of [7] can be adapted so as to give the Claim in the case $X, <_w$ has order type ω_1 but that it fails badly in the general case.

The result of 7.9 with MA in place of ($\mathcal{L}_2$) is due to Kunen [25] (see also [4]).

The result of 7.10 (and 7.13) first appeared in our note [62f] under the assumption that X is countably tight and Lindelöf in any of its finite powers.

The result of 7.14, stated in the form that every compact countably tight *ccc* space is separable, is due to Shapirovskii [70]. Note that the (L) part of 7.12 is an immediate consequence of this fact. Note also that the proof of 7.14 shows (without any additional axiomatic assumptions) that a countably tight space has precaliber $\aleph_1$ if and onlly if it is separable. This result is also due to Shapirovskii [96] (an explanation of this result can be found in [62o].

It turns out that (as in the case of (S) and (L)) the countability requirement in 7.14 is quite essential. To see this note that the spaces X_I and Y_I (when I and $\lambda \backslash I$ are both of size λ) from the proof of 0.10 are subspaces of

$$\{x \in \{0,1\}^A : |\{\alpha \in A : x(\alpha) = 1\}| \leq \kappa\}$$

and, therefore, both have tightness $\leq \kappa$. This means that the construction of 0.10 also produces a compact space X such that

$$d(X) > c(X) \cdot t(X).$$

This should be compared with the question 4 of [75; p. 78]. Note also that this shows that the Conjecture KH from page 49 of [75] is false. We conjecture that a refinement of the construction of 0.10 should be able to give us compact spaces X and Y such that

$$hd(X) > hL(X) \text{ and } hL(Y) > hd(Y).$$

The result of 7.15 first appeared in our note [62k]. Before that, the conclusion of 7.15 in the class of all *regular* spaces was deduced from the Conjecture (S) by Gardner and Pfeffer [15]. We refer the reader to [16] and [42] for more information of how this result fits into the area of topological measure theory.

Finally, we note that (S) is not a consequence of $MA_{\aleph_1}$ even in the class of all first countable spaces (see [49] and [0]). Similarly, $MA_{\aleph_1}$ doesn't resolve (L). This was first proved in [0] but it will also be a consequence of the result of §9.

8. PROPER FORCING
AXIOM AND PARTITIONS

The *ccc* property of partitions appears to be too restrictive to be really useful in considering the problems related to (S) and (L). The purpose of this section is the use the following standard forcing axiom ([4], [44]) in stating results about uncountable homogeneous sets.

(PFA) *If* P *is a proper poset and* $\{D_\alpha : \alpha < \omega_1\}$ *a family of dense subsets of* P, *then there is a filter* $\mathcal{G}$ *in* P *intersecting each* D_α.

Every *ccc* and every σ-closed poset is proper, and the composition of two proper posets is proper. These are the only facts we shall need for the first few results below. Clearly, $\text{MA}_{\aleph_1}$ is a consequence of PFA. Similarly to $\text{MA}_{\aleph_1}$, many of the interesting consequences of PFA have the form of Ramsey-type statements, so it seems natural to ask for such a Ramsey-type statement which is equivalent to PFA. At the moment such an equivalence is far from our reach since it clearly requires a much deeper understanding of the notion of proper poset than we presently have.

8.0. THEOREM. (PFA). *If* $[X]^2 = K_0 \cup K_1$ *is a given partition where* $X \subseteq \mathbf{R}$ *and where* K_0 *is open in* $[X]^2$, *then either there is an uncountable 0-homogeneous set, or else* X *is the union of countably many 1-homogeneous sets.*

PROOF. Here we identify $[X]^2$ with $\{\langle x, y \rangle : x < y \text{ in } X\}$ where the topology is induced from $\mathbf{R}^2$. Let P be the standard σ-closed poset which makes $\mathbf{R}$ of size $\aleph_1$. If X is not the union of $\aleph_0$ 1-homogeneous sets this is still true in the forcing extension of P. By considering the set-mapping

$$x \mapsto \overline{\{y : \langle x, y \rangle \notin K_0\}}$$

and applying 4.4 in $\mathbf{V}^P$ we get uncountable $\dot{Y} \subseteq X$ such that the poset $\dot{Q}$ of all finite 0-homogeneous subsets of $\dot{Y}$ is *ccc*. Now PFA for $P * \dot{Q}$ gives us uncountable 0-homogeneous subsets of X.

Let OCA denote the open coloring axiom of 8.0 and let us also consider the following closed set-mapping axiom.

(CSM) *If F is a closed set mapping on a set of reals, then either there is an uncountable F-free subset of $dom(F)$, or else F is the union of countably many connected functions.*

8.1. PROPOSITION. *OCA and CSM are equivalent statements.*

PROOF. We have already deduced OCA from CSM in the proof of 8.0. For the converse, identify F with the subset

$$X = \{\langle x, F(x)\rangle : x \in dom(F)\}$$

of the product $\mathbf{C} \times \exp(\mathbf{C})$ and note that freeness is an open relation on X.

Using the diagonalization statement 4.2 (in fact, its stronger version 4.3) instead of 4.4 in the above proof of 8.0 we get the following.

8.2. THEOREM. (PFA). *If X and Y are sets of reals of size $\aleph_1$, then there is a 1-1 map $f : X \to Y$ which is the union of countably many increasing subfunctions.*

By adding a little bit of MA to the statement of 8.2 we have the following well known fact ([2]).

8.3. COROLLARY. (PFA). *Every two $\aleph_1$-dense sets of reals X and Y are isomorphic.*

PROOF. For each two rational intervals I and J pick 1-1 maps

$$f_{IJ} : X \cap I \to Y \cap J \text{ and } g_{IJ} : Y \cap J \to X \cap I$$

which are the union of $\aleph_0$ increasing subfunctions $\{f_{IJ}^n\}$ and $\{g_{IJ}^m\}$, respectively. Let $\mathcal{Q}$ be the poset of all finite increasing mappings p from X into Y such that for every x in $dom(p)$ there exist rational intervals I and J and integers n and m such that either

$$p(x) = f_{IJ}^n(x) \text{ or } x = g_{IJ}^m(p(x)).$$

It is easily seen that $\mathcal{Q}$ is σ-centered and makes X and Y isomorphic.

Let us now give some applications of the open coloring axiom OCA of 8.0 just to give the reader a feeling about which kind of problems can be expected to be influenced by OCA.

8.4. PROPOSITION. (OCA).

(a) *Every uncountable subset of $\mathcal{P}(\omega)$ contains an uncountable chain or antichain.*

(b) *Every function f from an uncountable set of reals into the reals is monotonic on an uncountable set.*

(c) *If X and Y are two uncountable sets of reals then there is strictly increasing mapping from an uncountable subset of X into Y.*

(d) *Every uncountable Boolean algebra contains an uncountable antichain.*

PROOF. For (a), (b) and (c) just note that the inclusion on $P(\omega)$, when identified with $\{0,1\}^\omega$, is a closed relation and that strictly increasing is an open relation on the plane. If B is an uncountable algebra, first of all note that we may assume that B is a subalgebra of $P(\omega)$ and then use (a) and (b) (see [4; 6.14]).

The next result shows that OCA implies (G) of §3 in the class of all cometrizable space. We should note that its proof is just a reformulation of the proof of [18] using the terminology of OCA instead of PFA. It is given here only for the reader's convenience. Since 8.4(c) implies that the square of an uncountable subspace of the Sorgenfrey line has an uncountable discrete subspace, it follows that OCA also implies (M) of §3 in the class of all cometrizable spaces.

8.5. THEOREM. (OCA). *If X is a cometrizable space, then either X has a countable network, or an uncountable discrete subspace, or an uncountable subspace of the Sorgenfrey line.*

PROOF. We may assume X is a subset of $\{0,1\}^\omega$, that its topology is finer than that induced from $\{0,1\}^\omega$, and that for each x in X we have fixed a family B_x of closed subsets of $\{0,1\}^\omega$ whose restriction on X forms a local base of x in X. Let

$$Y = \{\langle x, F \rangle : x \in X, F \in B_x\}$$

and consider Y with the usual topology induced from the product of the Cantor set and its exponential space. We shall consider the following two open colorings of Y,

$$[Y]^2 = R_0 \cup R_1 \text{ and } [Y]^2 = L_0 \cup L_1$$

defined by the rules

$$\{\langle x, F \rangle, \langle y, G \rangle\} \in R_0 \text{ iff } x < y \text{ and } y \notin F$$

$$\{\langle x, F \rangle, \langle y, G \rangle\} \in L_0 \text{ iff } x < y \text{ and } x \notin G.$$

By OCA we have the following two cases

Case 1. Y is the union of countably many R_1-homogeneous sets $\{Y_{ri}\}_i$ and also Y is the union of countably many L_1-homogeneous sets $\{Y_{\ell i}\}_i$. For $i < \omega$, set

$$X_{ri} = \{x \in X : \exists F \ \langle x, F \rangle \in Y_{ri}\}, \text{ and}$$

$$X_{\ell j} = \{x \in X : \exists \, F \, \langle x, F \rangle \in Y_{\ell j}\}.$$

Then it is easily seen that $\{X_{ri} \cap X_{\ell j}\}_{ij}$ forms a countable network for X.

Case 2. Case 1 doesn't hold. By symmetry we may assume to have an uncountable $X_0 \subseteq X$ and for each x in X_0 an F_x in $\mathcal{B}_x$ such that

$$[\{\langle x, F_x \rangle : x \in X_0\}]^2 \subseteq R_0$$

Consider now the restriction of the partition $\langle L_0, L_1 \rangle$ on

$$Y_0 = \{\langle x, F \rangle : x \in X_0, F \in \mathcal{B}_x\}$$

and applying OCA we either have an uncountable 0-homogeneous set which would clearly be discrete subspace of X, or else Y_0 is the union of countably many 1-homogeneous sets $\{Y_{0i}\}_i$. Again for $i < \omega$, we set

$$X_{0,i} = \{x \in X_0 : \exists \, F \, \langle x, F \rangle \in Y_{0,i}\}.$$

By 8.4(b) there is uncountable $X_1 \subseteq X_0$ such that for all i, $X_{0i} \cap X_1$ is clopen in Cantor topology on X_1. It is now straightforward to check that the topology on X_1 induced from X is the same as the left-facing Sorgenfrey topology on X_1. This completes the proof.

The Ramsey-type statement OCA of 8.0 is also relevant to the existence of gaps in ω^ω, $<^*$ (or $[\omega]^\omega$, $\subset^*$). In fact, it implies that (ω_1, ω_1^*) is the only type of a (κ, λ) gap in ω^ω (or in $[\omega]^\omega$) for κ and λ regular and uncountable. This will be seen in the next proof.

8.6. THEOREM. (PFA). *The only types of gaps in ω^ω, $<^*$ are the following:* (ω_1, ω_1^*), (ω_2, ω^*) *and* (ω, ω_2^*).

PROOF. The only use of PFA is to make the continuum equal to $\aleph_2$ (see 62*l*). The rest of the result will be deduced solely from OCA. Suppose $\langle A, B \rangle$ is a gap in ω^ω, $<^*$ which has none of these three types. First of all, note that $\langle A, B \rangle$ cannot be an (ω_1, ω^*) or (ω, ω_1^*) gap. This follows from Rothberger's characterization of such gaps ([38]) and the fact that the partition (5) of §7 is open. Thus $A, <^*$ and $B, >^*$ have uncountable regular types and one of them, say, A has size $> \aleph_1$. By shrinking A we may assume that for every f in A there is cofinal $B_f \subseteq B$ such that for some fixed n_0,

$$\Gamma(f, g) = n_0 \text{ for all } f \text{ in } A \text{ and } g \text{ in } B_f.$$

Let $X = \{\langle f, g \rangle : f \in A \text{ and } g \in B_f\}$ and consider the partition

$$[X]^2 = K_0 \cup K_1$$

defined by:

$$\{\langle f, g \rangle, \langle \bar{f}, \bar{g} \rangle\} \in K_0 \text{ iff } \max\{\Gamma(f, \bar{g}), \Gamma(\bar{f}, g)\} > n_0.$$

Clearly, K_0 is open in $[X]^2$ and the argument from the proof of 3.7 shows that X is not the union of countably many 1-homogeneous sets. Thus there is a 0-homogeneous set $Y \subseteq X$ of size $\aleph_1$. We may assume that Y is of the form $\{\langle f_\alpha, g_\alpha \rangle : \alpha < \omega_1\}$ where f_α's are $<^*$-increasing. Note that g_α's must be different and, therefore, we may assume they are $<^*$-decreasing. Pick an $\bar{f}$ in A above f_α's. Then we can find an integer $n_1 \geq n_0$, uncountable $I \subseteq \omega_1$ and $s, t \in \omega^{n_1}$ such that for all α in I,

$$\Gamma(f_\alpha, \bar{f}), \Gamma(\bar{f}, g_\alpha) \leq n_1, s \subset f_\alpha \text{ and } t \subset g_\alpha.$$

Note that this means that $\Gamma(f_\alpha, g_\beta) \leq n_0$ for all α and β in I. But this contradicts the fact that Y is 0-homogeneous. The rest of the result follows from 3.10, Rothberger's characterization of the minimal cardinal of an unbounded subset of ω^ω, $<^*$ ([38], [8]), and the classical result of Hausdorff [20].

For f in ω^ω, let $U_f = \{\langle i, j \rangle : j \leq f(i)\}$. A *coherent family of functions indexed by a subset* I *of* ω^ω is a family $\{g_f : f \in I\}$ such that

(a) $g_f : U_f \rightarrow \omega$ for all f in I,
(b) $\{x \in U_f \cap U_h : g_f(x) \neq g_h(x)\}$ is finite for all f and h in I.

A coherent family $\{g_f : f \in I\}$ is *trivial* if there is a $g : \omega \times \omega \rightarrow \omega$ such that

$$g | U_f =^* g_f \quad \text{for all } f \text{ in } I.$$

8.7. THEOREM. (OCA). *Every coherent family of functions indexed by ω^ω is trivial.*

PROOF. Fix such a family $\{g_f : f \in \omega^\omega\}$ and define

$$[\omega^\omega]^2 = K_0 \cup K_1$$

by letting $\langle f, h \rangle$ in K_0 iff there is an x in $U_f \cap U_h$ such that $g_f(x)$ is different from $g_h(x)$. If we identify an f with the graph of the function g_f the natural topology of $P(\omega^3)$ induces a new separable metric topology on ω^ω which has the property that K_0 is open in its square. An argument very similar to that from the previous proof and the fact that every subset of ω^ω of size $\aleph_1$ is bounded (see 8.6) show that there are no uncountable 0-homogeneous sets. Thus, ω^ω is the union of countably many 1-homogeneous sets. So pick one such a set I which, moveover, has the property that for every f in ω^ω there is an h in I such that $f <^* h$. Define

$$g : \omega \times \omega \rightarrow \omega$$

by letting $g(i, j) = g_f(i, j)$, where f is any member of I such that $\langle i, j \rangle$ is in U_f; if such an f doesn't exist, let $g(i, j) = 0$. Then it is easily checked that g is well-defined and that it trivializes the family $\{g_f : f \in \omega^\omega\}$. This completes the proof.

The technique of "freezing" in the context of PFA and the reals is quite useful indeed. So, let us present it in more detail using a PFA-reformulation of a result of Kunen [25] (where the technique of "freezing" originated) as an example.

8.8. THEOREM. (PFA). $P(\omega_1)$ *is not a subalgebra of* $P(\omega)/FIN$ *and there is no strictly increasing map from* 2^{ω_1}, $<_{lex}$ *into* ω^ω, $<^*$ *(nor into* $[\omega]^\omega$, $\subset^*$*).*

PROOF. Suppose $\{f_\alpha : \alpha < \kappa\}$ and $\{g_\alpha : \alpha < \kappa\}$ form a (κ, κ^*) gap in $\omega^\omega, <^*$, where κ is a regular uncountable cardinal. We may assume that

(a) $\forall \alpha \; \forall n \; f_\alpha(n) \le g_\alpha(n)$

Let $X = \{\langle f_\alpha, g_\alpha \rangle : \alpha < \kappa\}$ and define partition

$$[X]^2 = K_0 \cup K_1$$

by letting $\{\langle f_\alpha, g_\alpha \rangle, \langle f_\beta, g_\beta \rangle\}$ into K_0 iff

$$\Gamma(f_\alpha, g_\beta) > 0 \text{ or } \Gamma(f_\beta, g_\alpha) > 0.$$

The following fact about such a kind of partition is of independent interest.

CLAIM. *The poset of all finite 0-homogeneous sets of the partition (1) is* κcc.

PROOF. Let $\{H_\xi : \xi < \kappa\}$ be given sequence of finite 0-homogeneous sets of (1). We may assume H_ξ are disjoint and increasing. For $\xi < \kappa$, let α_ξ be the minimum of H_ξ. By refining the sequence, we may assume to have an $m < \omega$ and s and t in ω^m such that for all ξ:

(b) $\forall n \ge m \; \forall \alpha \in H_\xi \; f_{\alpha_\xi}(n) \le f_\alpha(n)$,

(c) $\forall n \ge m \; \forall \beta \in H_\xi \; g_\beta(n) \le g_{\alpha_\xi}(n)$,

(d) $s \subset f_{\alpha_\xi}$ and $t \subset g_{\alpha_\xi}$.

Since $\{f_{\alpha_\xi} : \xi < \kappa\}$ and $\{g_{\alpha_\xi} : \xi < \kappa\}$ also form a gap there must be $\xi < \eta$ such that

$$\Gamma(f_{\alpha_\xi}, g_{\alpha_\eta}) > 0.$$

By (a) and (d), this means that for some $n \ge m$,

(e) $f_{\alpha_\xi}(n) > g_{\alpha_\eta}(n)$.

Using (b) and (c) this gives

(f) $\forall \alpha \in H_\xi \; \forall \beta \in H_\eta \; f_\alpha(n) > f_\beta(n)$

which shows that $H_\xi \cup H_\eta$ is 0-homogeneous. This proves the Claim.

Now we are ready for the proof of 8.8. Since proofs of all three statements are almost identical, we prove only that 2^{ω_1}, $<_{lex}$ is not embeddable into ω^ω, $<^*$. Otherwise, fix a strictly increasing map $\varphi : 2^{\omega_1} \to \omega^\omega$. For each h in 2^{ω_1} with $\aleph_1$ 0's and $\aleph_1$ 1's we fix the increasing enumerations

$$\{\alpha_\xi(h) : \xi < \omega_1\} = h^{-1}(1) \text{ and } \{\beta_\xi(h) : \xi < \omega_1\} = h^{-1}(0).$$

For $\xi < \omega_1$ let $f_\xi(h)$ in 2^{ω_1} be equal to h below the ordinal $\alpha_\xi(h)$, equal to 0 at $\alpha_\xi(h)$ and equal to 1 above $\alpha_\xi(h)$. Similarly for $\xi < \omega_1$, let $g_\xi(h)$ in 2^{ω_1} be equal to h below the ordinal $\beta_\xi(h)$, equal to 1 at $\beta_\xi(h)$ and equal to 0 above $\beta_\xi(h)$.

Let P be the standard σ-closed poset for making $\mathbf{R}$ of size $\aleph_1$. Working in $\mathbf{V}^P$ we see that for each h in 2^{ω_1} the functions $f_\xi(h)$ and $g_\xi(h)$ for $\xi < \omega_1$ are all in the ground model. So $\varphi(f_\xi(h))$ and $\varphi(g_\xi(h))$ are all defined. Moreover, $\varphi(f_\xi(h))$'s are $<^*$-increasing while $\varphi(g_\xi(h))$'s are $<^*$-decreasing. By the cardinality considerations (in $\mathbf{V}^P$) there must be an $\dot{h}$ in 2^{ω_1} with $\aleph_1$ 0's and $\aleph_1$ 1's for which $\varphi(f_\xi(\dot{h}))$'s and $\varphi(g_\xi(\dot{h}))$'s form a gap in ω^ω, $<^*$. The Claim gives us a ccc poset $\dot{Q}$ which forces an uncountable 0-homogeneous set for the associated partition, i.e., freezes the gap. Returning to $\mathbf{V}$ and forcing internally with $P * \dot{Q}$, we get an h in 2^{ω_1} with $\aleph_1$ 0's and $\aleph_1$ 1's, an uncountable set $I \subseteq \omega_1$, and integer n_0 such that for all $\xi < \eta$ in I:

(2) $\Gamma(\varphi(f_\xi(h)), \varphi(g_\xi(h))) = n_0$,
(3) $\Gamma(\varphi(f_\xi(h)), \varphi(g_\eta(h))) > n_0$.

Then as in the proof of 8.6 we conclude that $\varphi(f_\xi(h))$'s and $\varphi(g_\xi(h))$'s for ξ in I form a gap, so $\varphi(h)$ has no place in ω^ω. This finishes the proof.

The open coloring axiom for the reals in 8.0 can be strengthened in the following fashion:

(OCA) *If X is a regular space with no uncountable discrete subspace and if $[X]^2 = K_0 \cup K_1$ is a given partition with K_0 open, then either there is an uncountable 0-homogeneous set, or else X is the union of countably many 1-homogeneous sets.*

In the general case the closed set mapping axiom becomes:

(CSM) *If X is a regular space with no uncountable discrete subspace and if F maps X into the family of all closed subsets of X, then either there is an uncountable F-tree subset of X or else F is the union of countably many connected subfunctions.*

Clearly, CSM implies OCA but the proof of their equivalence in the case of sets of reals doesn't work in the general case since we can no longer assert that $X \times \exp(X)$ satisfies the hypothesis of OCA. It should also be clear that CSM implies both (S) and (L). The argument of 8.5 can be used to show that CSM implies both (T) and (G) of §3 in the class of all first countable spaces. In this context, another axiom (suggested to us by D. Fremlin) worth considering might be the following:

(OSM) *If X is a regular space with no uncountable discrete subspace and if for each x in X, U_x is an open set containing x, then X is the union of countably many sets X_n such that for all n and x and y in X_n either $x \in U_y$ or $y \in U_x$.*

Let us now return to the Conjecture (S) and present a method of introducing uncountable homogeneous sets in a situation where the coloring is not
that simple as in OCA. Recall that a poset P is *proper* ([4], [5], [44]) if for
any large enough regular cardinal θ and a countable elementary submodel
M of $\mathbf{H}_\theta$ containing P every condition p from $P \cap M$ can be extended to
a M-P-generic condition q. Recall also that q is M-P-generic if for every
$r \leq q$ and dense open subset D of P which is a member of M there is an s
in $D \cap M$ compatible with r. Our posets will usually be sets of certain pairs
$p = \langle D_p, N_p \rangle$, where D_p is the working part of the condition (usually a finite
set or a function with finite domain) and N_p is a finite $\in$-chain of countable
elementary submodels of some large enough structure $\mathbf{H}_\lambda$. We shall always
have the requirement that N_p *separates* D_p with meaning that for every two
distinct x and y in D_p there is an N in N_p containing exactly one of them.
Note that this uniquely determines a linear ordering $<_p$ on D_p:

$$x <_p y \text{ iff } \exists N \in N_p \ (x \in N \ \& \ y \notin N).$$

The crucial part of any of our definitions will be the "side conditions" deter-
mining how D_p and N_p interact. The side conditions are usually obtained by
a fine analysis of all the reasons which prevent the working part of a given
condition to be generic over a given countable elementary submodel. The
models of N_p are there to mimic the actual countable elementary submodel
M (of a much larger structure than $\mathbf{H}_\lambda$) over which we would like p to be
generic.

8.9. THEOREM. (PFA). *The Conjecture (S) is true.*

PROOF. Let $X, <$ be a right separated regular space of type ω_1 satisfying
the hypothesis of (S). We may assume $X, <$ is a member of the structure $\mathbf{H}_{\aleph_2}$.
For each x in X we fix a neighborhood U_x whose closure contains nothing
above x. Let P be the set of all pairs $p = \langle D_p, N_p \rangle$, where:

  (a) D_p is a finite subset of X such that $x \notin U_y$ for all $x < y$ in D_p.
  (b) N_p is a finite $\in$-chain of countable elementary submodels of $\mathbf{H}_{\aleph_2}$ con-
      taining $X, <$ and the sequence of U_x.
  (c) For all $x < y$ in D_p there is an N in N_p such that x is in N but y is
      not.

The ordering on P is defined by

$$p \leq q \text{ iff } D_p \supseteq D_q \text{ and } N_p \supseteq N_p.$$

It should be clear that the following Claim finishes the proof of 8.8 since
every condition p for which D_p has at least two elements forces that the
generic discrete subspace of X is uncountable.

CLAIM. P *is proper.*

PROOF. Let M be a countable elementary submodel of $\mathbf{H}_\theta$ containing p, P and X, $<$, where θ is a large enough regular cardinal. Let

$$q = \langle D_p, N_p \cup \{M \cap \mathbf{H}_{\aleph_2}\} \rangle$$

It is clear that q is in P and that $q \le p$. We claim that q is M-P-generic. To see this pick $r \le q$ and dense open subset D of P in M. We may assume r is in D. Let $\bar{r} = r \cap M$, $D = D_r \setminus M$ and $n = |D|$. We may assume $n \ge 1$. Let T be the set of all n element subsets E of X for which there exist $s \le \bar{r}$ in D such that $D_s \setminus D_{\bar{r}} = E$. Then clearly T is in M. We claim that it has the following property when we identify an E of T with the $<$-increasing sequence which enumerates it.

(d) The order type of T, $<_{lex}$ is equal to ω_1^n.

This is in fact a consequence of the following general fact about the ordinal ω_1^n which is proved by an obvious induction on n.

(e) Suppose $N_0 \in \cdots \in N_{n-1}$ is a chain of countable elementary submodels of $\mathbf{H}_{\aleph_2}$. Suppose N_0 contains a subset S of $^n\omega_1$ which has an element $\langle \alpha_0, \ldots, \alpha_{n-1} \rangle$ such that

$$N_0 \cap \omega_1 \le \alpha_0 < N_1 \cap \omega_1 \le \cdots < N_{n-1} \cap \omega_1 \le \alpha_{n-1}.$$

Then the type of $S <_{lex}$ is ω_1^n.

Again, a general fact about $S \subseteq {}^n\omega_1$ of type ω_1^n is that the tree $\hat{S}$ of initial parts of members of S contains an everywhere $\aleph_1$ splitting subtree. This is also proved by an obvious induction on n. Thus, we may suppose that our T is just the set of all top nodes of an everywhere $\aleph_1$ splitting subtree $\hat{T}$ of $[X]^{\le n}$.

Let

$$X_0 = \{x \in X : \{x\} \in \hat{T}\}.$$

Then X_0 is in M and there is a countable $Y_0 \subseteq X_0$ in M such that $\bar{Y}_0 \supseteq X_0$. By the choice of U_x's there must be an x_0 in Y_0 ($\subseteq M$) such that $x_0 \notin U_x$ for all x in D. Let

$$X_1 = \{x \in X : \{x_0, x\} \in \hat{T}\}.$$

Then working as above we can find an x_1 in $X_1 \cap M$ such that $x_1 \notin U_x$ for all x in D. Proceeding in this manner we produce an $E = \{x_0, \ldots, x_{n-1}\}$ in $T \cap M$ such that $x_i \notin U_x$ for all $i < n$ and x in D. Since E is in $T \cap M$, by the definition of T (and elementarity of M) there is an s in $D \cap M$ such that

$$s \le \bar{r} \text{ and } D_s \setminus D_{\bar{r}} = E.$$

It should be clear that

$$\langle D_s \cup D_r, N_s \cup N_r \rangle$$

is a condition of P. (The condition (a) is satisfied by our choice of the set E.) This shows that r is compatible with a member of $D \cap M$ and finishes the proof that P is proper.

Note that we have not used much of the regularity of the right separated space $X, <$ in the above proof of 8.9. Namely, suppose we have an uncountable space X (with no separation axioms assumed) with a well ordering $<$. In order to produce a proper poset $\mathcal{P}$ which produces an uncountable discrete subspace of X all we need is to have for every x in X a neighborhood U_x such that

$$y \notin \bar{U}_x \text{ for all } y > x.$$

This enables us to state the following result which is slightly stronger than 8.9.

8.10. THEOREM. (PFA). *Let X be a topological space (with no separation axiom assumed) with no uncountable discrete subspace. Then for every open cover $\mathcal{U}$ of X there is a countable $\mathcal{U}_0 \subseteq \mathcal{U}$ such that X is covered by the closures of members of $\mathcal{U}_0$.*

8.11. COROLLARY. (PFA). *Let X be a Hausdorff space with no uncountable discrete subspaces. Then every point of X is the intersection of countably many open sets.*

PROOF. Fix x_0 in X and for every $x \neq x_0$ in X a neighborhood U_x whose closure doesn't contain x_0. Now apply 8.10 to $X \setminus \{x_0\}$ and

$$\mathcal{U} = \{U_x : x \in X \setminus \{x_0\}\}.$$

8.12. COROLLARY. (PFA). *Every Hausdorff space with no uncountable discrete subspace has size at most $2^{\aleph_0}$.*

PROOF. This follows from 8.11 using a well-known result from the theory of cardinal functions in topology (see [71; 2.15(a)]).

We shall now present two more uses of the same method where the elementary submodels as side conditions are used in a more essential way than in the proof of 8.9. First we need to introduce a definition.
A sequence $\langle A_\alpha : \alpha < \omega_1 \rangle$ of subsets of ω_1 is *coherent* iff

(a) $A_\alpha \subseteq \alpha$ for all α, and
(b) $A_\alpha =^* A_\beta \cap \alpha$ for $\alpha < \beta < \omega_1$.

In [54] we have proved that for any coherent sequence A_α there is an uncountable set $A \subseteq \omega_1$ such that either $A \cap \alpha \subseteq^* A_\alpha$ for all $\alpha < \omega_1$, or else $A \cap A_\alpha$ is finite for all $\alpha < \omega_1$. This was deduced from a partition property which also implied the Conjecture (S). We shall now see that the methods of [54], [55], and [57] give us a much stronger result in this direction. The question appeared in an earlier version of Dow's paper [9] and it originally came from his interest in the space $SU(\omega_1)$ of subuniform ultrafilters on ω_1, i.e., free ultrafilters on ω_1 concentrating on countable subsets of ω_1. Namely, note that there is a natural correspondence between clopen subsets of $SU(\omega_1)$

and coherent sequences on ω_1. In the terminology of $SU(\omega_1)$ our result will say that there is a uniform ultrafilter $\mathcal{U}$ on ω_1 (i.e., a point of $U(\omega_1)$) such that for every clopen set C in the space $SU(\omega_1)$, the point $\mathcal{U}$ of $\beta\omega_1 \setminus \omega_1$ has a neighborhood which is either contained in the closure of C in $\beta\omega_1 \setminus \omega_1$, or it is disjoint from it. So, in some sense, $\mathcal{U}$ is also an ultrafilter in the algebra of all clopen subsets of $SU(\omega_1)$.

8.13. THEOREM. (PFA). *There is a uniform ultrafilter $\mathcal{U}$ on ω_1 such that for any coherent sequence A_α there is an A in $\mathcal{U}$ such that either $A \cap \alpha \subset^* A_\alpha$ for all α, or else $A \cap A_\alpha =^* \varnothing$ for all α.*

PROOF. Since there exist only $\aleph_2$ coherent sequences (see [62ℓ]), we shall construct $\mathcal{U}$ recursively in ω_2 steps. The result is proved if we show how to handle the following situation:

We have a filter base $\mathcal{F} \supset \{\omega_1 \setminus \alpha : \alpha < \omega_1\}$ of size $\aleph_1$ and a coherent sequence $\{A_\alpha : \alpha < \omega_1\}$. We have to produce an $A \subseteq \omega_1$ such that:

  (a) $A \cap B \neq \varnothing$ for all B in $\mathcal{F}$.
  (b) $A \cap \alpha \subset^* A_\alpha$ for all α, or else $A \cap A_\alpha =^* \varnothing$ for all α.

Case 1. For any $B \in \mathcal{F}$ and any $\{B_n : n < \omega\} \subseteq P(\omega_1)$ such that $B_n \cap A_\alpha$ is finite for all $n < \omega$ and $\alpha < \omega_1$, the set B minus $\bigcup_n B_n$ is uncountable.

Let P_1 be the set of all $p = \langle A_p, \mathcal{N}_p \rangle$ such that:

  (1) A_p is a finite subset of ω_1.
  (2) $\mathcal{N}_p$ is a finite $\in$–chain of countable elementary submodels of $\mathbf{H}_{\aleph_2}$ containing everything relevant.
  (3) For all $\alpha < \beta$ in A_p there is an N in $\mathcal{N}_p$ which contains α but not β.
  (4) For all N in $\mathcal{N}_p$, α in $A_p \setminus N$, and C in $N \cap P(\omega_1)$ containing α there is a β in ω_1 such that $C \cap A_\beta$ is infinite.

We order P_1 by letting $q \leq p$ iff:

  (5) $A_q \supseteq A_p$,
  (6) $\mathcal{N}_q \supseteq \mathcal{N}_p$,
  (7) For all β in A_p and α in $A_q \setminus A_p$, $\alpha < \beta$ implies α is in A_β.

CLAIM. P_1 *is proper.*

PROOF. Let M be a countable elementary submodel of $\mathbf{H}_\theta$ containing everything relevant, where θ is a large enough regular cardinal. Pick a p in $P_1 \cap M$ and let

$$q = \langle A_p, \mathcal{N}_p \cup \{M \cap \mathbf{H}_{\aleph_2}\}\rangle.$$

Clearly, q is a condition $\leq p$. We claim that q is M-P_1-generic.

So fix an $r \leq q$ and a dense open subset $\mathcal{D}$ of P_1 which is a member of M. We may assume that r is in $\mathcal{D}$. Let $\bar{r} = r \cap M$ and $A = A_r \setminus M$. Let $n = |A|$. We may assume $n \geq 1$. Call a subset X of ω_1 *good* iff every endsection of X has infinite intersection with some A_β. So in particular X has to be uncountable. Working as in the proof of 8.9 one shows that there is an everywhere $\aleph_1$-splitting tree $T \subseteq [\omega_1]^{\leq n}$ of height $n + 1$ such that:

(8) For every nonterminal t in T, $X_t = \{\xi < \omega_1 : t \cup \{\xi\} \in T\}$ is good.

(9) For every t in $T \cap [\omega_1]^n$ there is a condition r_t in $\mathcal{D}$ which extends $\bar{r}$ such that $t = A_{r_t} \setminus A_{\bar{r}}$.

Let $\delta = M \cap \omega_1$ and $\gamma = \min A$. Fix an $\bar{\alpha} < \delta$ above the maximum of $A_{\bar{r}}$ such that

$$A_\beta \cap [\bar{\alpha}, \delta) = A_\gamma \cap [\bar{\alpha}, \delta) \text{ for all } \beta \text{ in } A.$$

Since $X_\varnothing$ is in M and since it satisfies (8), there must be an α_0 in $[\bar{\alpha}, \delta)$ such that α_0 is also in the intersection of $X_\varnothing$ and A_γ. Similarly one finds α_1 in

$$X_{\langle \alpha_0 \rangle} \cap A_\gamma \cap [\alpha_0, \delta),$$

and so on. Proceeding in this way we construct a $t = \{\alpha_0, \dots, \alpha_{n-1}\}$ in $T \cap M$ such that $t \subseteq A_\gamma$. Then it follows that r and r_t (which is a member of $M \cap \mathcal{D}$) are two compatible elements of $\mathcal{P}_1$ which was to be proved.

For $\xi < \omega_1$ and B in $\mathcal{F}$, set

$$\mathcal{D}_{\xi, B} = \{p \in \mathcal{P}_1 : (A_p \cap B) \setminus \xi \neq \varnothing\}.$$

CLAIM. $\mathcal{D}_{\xi, B}$ is dense open in $\mathcal{P}_1$ for all ξ and B.

PROOF. This follows from our assumption of Case 1.

It should be now clear that any filter of $\mathcal{P}_1$ intersecting all $\mathcal{D}_{\xi, B}$'s produces an A with properties (a) and (b) above.

Case 2. There exist B in $\mathcal{F}$ and $\{B_n : n < \omega\} \subseteq P(\omega_1)$ such that:

(d) B minus $\bigcup_n B_n$ is countable.

(e) $B_n \cap A_\alpha$ is finite for all n and all α.

Note that this means that there is no uncountable $C \subseteq B$ such that $C \cap \alpha \subseteq^* A_\alpha$ for all $\alpha < \omega_1$.

Let $\mathcal{P}_2$ be the set of all finite subsets of B. We order $\mathcal{P}_2$ by letting $q \leq p$ iff

(10) $q \supseteq p$,

(11) For all β in p and α in $q \setminus p$, $\alpha < \beta$ implies α is not in A_β.

CLAIM. $\mathcal{P}_2$ is a ccc poset.

PROOF. First of all, note that the following is an immediate consequence of the argument from the Case 1 (see also [54; §2]).

Subclaim. (PFA). *For every coherent sequence* $\{B_\alpha : \alpha < \omega_1\}$ *on* ω_1 *there is an uncountable* $C \subseteq \omega_1$ *such that either* $C \cap \alpha \subseteq^* B_\alpha$ *for all* α, *or else* $C \cap B_\alpha$ *is finite for all* α.

So, let $\{F_\xi : \xi < \omega_1\}$ be a sequence of elements of $\mathcal{P}_2$. We may assume F_ξ's are disjoint and of the same size $n \geq 1$. By applying the Subclaim in n successive steps, we can find uncountable $X \subseteq \omega_1$ such that for all $i < n$, if

$$D_i = \{\alpha : \exists \, \xi \in X \, (\alpha \text{ is the } i \text{th element of } F_\xi)\},$$

then $D_i \cap A_\alpha$ is finite for all α. This means that if ξ in X has the property that there exist infinitely many F_η's for η in X below the $\min F_\xi$, then F_ξ must be compatible, in P_2, with one of them. This completes the proof.

For $\xi < \omega_1$ and $C \in \mathcal{F}$, set

$$\mathcal{E}_{\xi,C} = \{p \in P_2 : (p \cap C) \setminus \xi \neq \varnothing\}$$

CLAIM. $\mathcal{E}_{\xi,C}$ is dense open for all ξ and C.

PROOF. $C \cap B$ is uncountable for all C in $\mathcal{F}$.

Now any filter of P_2 intersecting all $\mathcal{E}_{\xi,C}$'s produces an $A \subseteq B$ satisfying the conditions (a) and (b) above. This completes the proof.

8.14. THEOREM. (PFA). *Suppose X is a regular space and that Y is a non-closed subset of X. Then either X contains an uncountable free sequence which is a subset of Y, or else Y contains an infinite subset with all accumulation points outside of Y.*

PROOF. Fix an x_0 in $\bar{Y} \setminus Y$ and a maximal filter $\mathcal{F}$ on Y generated by sets which are closed relative to Y and whose closure in X contains x_0. Fix also a large enough regular cardinal λ such that $\mathbf{H}_\lambda$ contains x_0, X, Y and $\mathcal{F}$. Let P_3 be the set of all pairs $p = \langle U_p, \mathcal{N}_p \rangle$, where

(a) $\mathcal{N}_p$ is a finite $\in$-chain of countable elementary submodels of $\mathbf{H}_\lambda$ containing x_0, X, Y and $\mathcal{F}$.

(b) U_p is a finite function from a subset D_p of Y separated by $\mathcal{N}_p$ such that $U_p(x)$ is an open neighborhood of x in X whose closure doesn't contain x_0 and which belongs to any model of $\mathcal{N}_p$ containing x. Moreover, the separation is mutual in the sense that for every two models in $\mathcal{N}_p \cup \{\mathbf{H}_\lambda\}$ there is a point in D_p which is in the bigger but not in the smaller one.

(c) If $x <_p y$ are in D_p, then x is in $U_p(y)$. (The definition of the ordering $<_p$ can be found just before the statement of 8.9.)

(d) If N is in $\mathcal{N}_p$ and if x is the minimal point of D_p not in N, then x is in the closure of $Z \cap N$ for all Z in $\mathcal{F} \cap N$.

For p and q in P_3, set $p \leq q$ iff

$$U_p \supseteq U_q \text{ and } \mathcal{N}_p \supseteq \mathcal{N}_q.$$

CLAIM. *If the conclusion of 8.14 fails, P_3 is proper.*

PROOF. Let M be a countable elementary submodel of $\mathbf{H}_\theta$ containing everything relevant, where θ is a large enough regular cardinal. Pick a p in $P_3 \cap M$ and set

$$q = \langle D_p, \mathcal{N}_p \cup \{M \cap \mathbf{H}_\lambda\} \rangle.$$

Then q is in P_3 and $q \leq p$. We claim that q is M-P_3-generic. To see this, pick $r \leq q$ and a dense open subset $\mathcal{D}$ of P_3 from M. Since we have to find a condition from $\mathcal{D} \cap M$ compatible with r, we may assume r is in $\mathcal{D}$. Let $\bar{r} = r \cap M$, $D = D_r \setminus M$ and $n = |D|$. We may assume $n \geq 1$. Let T be the set of all n element subsets E of Y for which there exists $s \leq \bar{r}$ in $\mathcal{D}$ such that $D_s \setminus D_r = E$. Note that T is an element of M. Let $\bar{T}$ be the set of all subsets of elements of T. Then working as in 8.9 and 8.13 we can find (in M) a subposet S of $\bar{T}$ of height $n + 1$ containing the empty set such that for every nonmaximal F in S,

(e) The closure of $S_F = \{y \in Y : F \cup \{y\} \in S\}$ relative to Y is in $\mathcal{F}$.

Let z_0 be the $<_r$-minimal point of D, and let

$$U = \bigcap_{y \in D} U_r(y).$$

It suffices to find a maximal element of $S \cap M$ which is a subset of U. Note that by (c), z_0 is an element of U. By the property (e) of the poset $S, \subseteq$ and the fact that $\varnothing$ is in S, we know that $S_\varnothing$ is an element of M whose closure relative to Y, call it Z, is in $\mathcal{F}$. By (d), we can pick an element z from $Z \cap M \cap U$. Since the conclusion of 8.14 fails, Y is a countably tight space ([95], [69], [71]). So working in M, we can find a countable subset Y_0 of $S_\varnothing$ whose closure contains z. This means that we can pick y_0 in $Y_0 \cap U$. Since Y_0 is countable, it is a subset of M. Hence y_0 is an element of M, so we can now consider $S_{\{y_0\}}$, and so on. Proceeding in this way we get as before a condition s in $\mathcal{D} \cap M$ compatible with r. This finishes the proof.

For $\alpha < \omega_1$, set

$$\mathcal{D}_\alpha = \{p \in \mathcal{P}_3 : \exists N \in \mathcal{N}_p \ (\alpha \in N \ \& \ D_p \setminus N \neq \varnothing)\}.$$

CLAIM. *If Y contains no infinite subset with all accumulation points in $X \setminus Y$, $\mathcal{D}_\alpha$ is dense open for all α.*

PROOF. Pick q in $\mathcal{P}_3$ and $\alpha < \omega_1$. Let N be a countable elementary submodel of $\mathbf{H}_\lambda$ containing $q, \alpha, X, Y, \mathcal{F}$ and x_0. Let F be the intersection of all sets of the form $\overline{Z \cap N}$, where Z is an element of $\mathcal{F} \cap N$ whose closure contains x_0. By our assumption about Y, $F \cap Y$ is nonempty. So pick a y in $F \cap Y$ and an open set $U_p(y)$ of X containing $D_q \cup \{y\}$ whose closure doesn't contain x_0. Let $U_p \mid D_q$ be equal to U_q, and let $\mathcal{N}_p$ be equal to $\mathcal{N}_q \cup \{N\}$. Then it is easily checked that $p = \langle U_p, \mathcal{N}_p \rangle$ satisfies the requirements (a)-(d) for being a member of $\mathcal{P}_3$. Note that p is in $\mathcal{D}_\alpha$ and $p \leq q$. This finishes the proof.

It is clear that a filter of $\mathcal{P}_3$ intersection all the $\mathcal{D}_\alpha$'s will give us an uncountable free sequence in X. This completes the proof.

8.15. COROLLARY. (PFA). *If X is a countably tight compact space and if Y is a non-closed subset of X, then there is a sequence of elements of Y converging to a point outside Y.*

PROOF. By 8.14, Y contains an infinite countable subset Y_0 with all accumulation points outside Y. Let Z be a minimal subset of X containing Y_0 which contains an accumulation point of every one of its infinite subsets. Then Z has size continuum and must be closed in X by 8.14. Thus $Z = \bar{Y}_0$. Since the continuum is equal to $\aleph_2$ (see [62ℓ]), by the Cech-Pospisil argument (see [71]) there must be a z in $Z \setminus Y_0$ with a local basis in Z of size smaller or equal to $\aleph_1$. Now a standard use of $MA_{\aleph_1}$ produces a sequence of elements of Y_0 converging to z.

8.16. COROLLARY. (PFA). *Every countably tight compact space has a point of countable character.*

PROOF. Let Y be a countably tight compact space and let P be the standard σ-closed poset for making Y of size at most $\aleph_1$. The topology of Y generates (as a basis) a possibly bigger topology on Y in V^P. We claim that Y with the new topology is still compact in V^P. Otherwise, working in V^P and considering a compactification X of Y we would get into the situation of 8.14 which would produce us a proper poset $\dot{P}_3$ which forces an uncountable free sequence in Y. Returning to V and forcing internally with $P * \dot{P}_3$ would give us an uncountable free sequence in Y, a contradiction. (We are using here the algebraic form of the free sequence; see [62o].) The Cech-Pospishil argument in V^P applied to the (small compact) space Y shows that Y contains a G_δ-point. Since this fact is clearly absolute between V and V^P, the proof is finished.

The above are just instances of a quite general approach in constructing proper partial orderings described in [55]. For example, 8.0, 8.2 and 8.8 can all be done also using this approach with proofs which might even be considered more natural and simpler. Let us illustrate this on 8.0. So let

$$[X]^2 = K_0 \cup K_1$$

be an open partition on a set of reals X with property that X is not the union of countably many 1-homogeneous sets. The poset P_4, which will force an uncountable 0-homogeneous set, will be the set of all pairs $p = \langle H_p, N_p \rangle$, where as usual:

(i) H_p is a finite 0-homogeneous subset of X.
(ii) N_p is a finite $\in$-chain of countable elementary submodels of $\mathbf{H}_{\aleph_3}$ containing X, K_0, K_1.
(iii) For all $x \neq y$ in H_p there is an N is N_p such that $N \cap \{x, y\}$ has exactly one element.
(iv) If x from H_p is not an element of N from N_p, then x is not in any 1-homogeneous subset of X which happens to be a member of N.

The ordering of P_4 is also as usual: $p \leq q$ iff $H_p \supseteq H_q$ and $N_p \supseteq N_q$. Then P_3 is a proper poset and a M-P_4-generic condition q below a p in $P_4 \cap M$ will again have the form

$$\langle H_p, N_p \cup \{M \cap \mathbf{H}_{\aleph_3}\}\rangle$$

We leave the proof of this to the interested reader.

It is reasonable to expect that this approach will find many other applications as our understanding of it, in the form as it is, increases. The recent "combinatorial" translations of this method [13; Lemma 1L] tend sometimes to hide trivial reasons for properness and therefore lower our ability of recognizing it. That is, we see no point in the obvious (but lengthy) computation (see [13]) how much closed the elementary submodels of the side condition $\mathcal{N}_p$ of p have to be in order to make the argument of proving properness of $\mathcal{P}$ work. The point should be in building new combinatorics which will enable us to construct *new* posets and discover *new* reasons for the existence of generic conditions. We should note that the present situation of our understanding this point is extremely unsatisfactory. For example, one can hardly find in today's literature two distinct combinatorial arguments which are finding a condition and proving its genericity with respect to a given countable elementary submodel.

None of the statements of this section needs large cardinals for its consistency. This may not be so clear for the last three statements since the side conditions obviously do collapse cardinals. In [57] we have presented an improvement of the side conditions which removes the need for large cardinals especially in situations where the forced object can be assumed to be a subset of ω_1. Namely, all we have to do is to improve the side condition

$$\mathcal{N}_p = \{N_0, \dots, N_k\}_\in$$

to have the more flexible form

$$\vec{\mathcal{N}}_p = \{\mathcal{N}_0, \dots, \mathcal{N}_k\},$$

where now $\mathcal{N}_i$ is no longer a single elementary submodel, but a finite set of countable elementary submodels with the same transitive collapse. The $\in$-chain requirement now becomes: If $i < j < k$, then for all N in $\mathcal{N}_i$ there is an M in $\mathcal{N}_j$ such that $N \in M$. The argument of proving properness for the modified version is usually only slightly different from the old one. The need for large cardinals is removed since if two conditions p and q have the same working parts (usually finite subsets of ω_1) and the same sequence of transitive collapses of members of their side conditions, then they are compatible. Thus, the modified poset satisfies a strong chain condition which can be iterated preserving cardinals.

8.17. Remarks.

Some basic and general facts about PFA can be found in [4], [5] and [44].

The main advance towards the Open Coloring Axiom for sets of reals was made by Baumgartner ([2], [3]). Its formulation, in a slightly weaker form than 8.0, was made by Abraham, Rubin and Shelah [1]. (The formulation of OCA in its present form was made by the author during the Spring of 1982 using the above proof with elementary submodels as side conditions.) The discovery of Baumgartner ([2], [3]) is that a CH-diagonalization argument can be used in building *ccc* poset which forces homogeneous sets for certain open partitions on sets of reals. One of the main contributions of Abraham, Rubin and Shelah [1] is the fact that this diagonalization (which, by the way, has length ω_1) can also be done in certain situations where CH fails. This enabled them to construct models of various forms of the Open Coloring Axiom where the continuum is $> \aleph_2$. That Baumgartner's argument can be done in ZFC and that, in fact, can be considered as very natural generalizations of the classical diagonalization arguments of Sierpinski-Zygmund [47] and Sierpinski [46] (see also [27; §35]), was noticed by the author in [58].

The result of 8.7 shows that OCA has some influence on calculating the strong homology of certain subsets of Euclidean spaces. An explanation of why this is so plus some other mathematical and historical facts can be found in [106] and [107].

The results of Kunen [25] have been first formulated using the PFA-terminology by Baumgartner [4]. Unfortunately, the result 8.8 appeared in [4] in a slightly weaker form. In [68], Woodin showed that in the Kunen model ($MA_{\aleph_1}$ plus the conclusion of 8.8) every homomorphism from C[0, 1] into a commutative Banach algebra is continuous. Here C[0, 1] denotes the algebra of all continuous real functions on the interval [0, 1] with the supremum norm.

The result of 8.9 is a formulation of the result of [54] using the PFA terminology.

The result of 8.13 first appeared in our note [62i].

Let us give a brief description of our first formulation of 8.9 as it appeared in [91]. We started with a regular space X which contains a right separated subspace $Y, <$ of type ω_1. For each y in Y, we picked two open neighborhoods U_y and V_y of y in X containing nothing in Y above y such that the closure of U_y is contained in V_y. We assumed that Y and $\{U_y\}$ satisfy the following combinatorial property:

(s) For every countable elementary submodel M of $\mathbf{H}_{\aleph_2}$ containing everything relevant, for every uncountable Z in $P(Y) \cap M$, and every finite $D \subseteq Y \setminus M$,

$$Z \cap (\bigcap_{y \in D} U_y) \neq \varnothing.$$

Then assuming Y is hereditarily separable, we produced proper poset P_3 which forced uncountable subset Z of Y such that for all y in Z,

$$\{x \in Z : x \leq y\} \subseteq U_y.$$

Note that this means that Z_0 is an uncountable free sequence in X. In 8.14 we have reformulated that argument using the following translations: The role of hereditary separability of Y is replaced by the countable tightness of Y. The local countability of Y and the combinatorial property (s) is replaced by the fact that if Z is the countable family of subsets of Y, then the intersection of the closures of members of Z intersects Y. In order to make this second replacement work it is essential to have elementary submodels as side conditions rather than closed and unbounded subsets of ω_1, as it was the case in [91]. Note that the side condition (d) from the definition of P_3 is made just to provide an analogue of (s) in the situation of the proof of 8.14.

The result of 8.15 is due to Balogh [92], Fremlin [93] and Nyikos. We have formulated 8.14 in its present form in order to (among other things) have 8.15 as a consequence. We should note that Balogh, Fremlin and Nyikos have many other interesting results of this kind which are frequently stated using the terminology of a stronger forcing axiom than PFA. We believe that the reason for this is that they needed the continuum to be equal to $\aleph_2$ and the result of [62ℓ] wasn't available at the time they were working on the subject. Finally, we note that the results of Balogh, Fremlin and Nyikos were proved consistent just relative to the consistency of ZF by Dow [94] by replacing the side conditions in the manner described just before the Remark 8.17. The result of 8.16 is also due to Dow [94].

As indicated in [54; §2] our interest in coherent sequences on ω_1 came from an unpublished example of F. Galvin concerning the tightness of the product of two spaces of size $\aleph_1$ with unique non-isolated points. A space X of size $\aleph_1$ with unique non-isolated point $*$ is usually given on the set $\omega_1 \cup \{*\}$ where the points of ω_1 are isolated and where the neighborhoods of $*$ are given by a proper ideal on ω_1 containing all singletons. It is known (see [75; §3]) that there could be two such spaces which are both countably tight but whose product is not countably tight. We would like to point out here that a proof almost identical to that of the Subclaim from Case 2 of 8.13 (or Theorem 6 of [54]) gives the following interesting fact under the assumption of PFA:

> (Q) *Suppose X and Y are Lindelöf countably tight spaces of size $\aleph_1$ and that both X and Y have exactly one non-isolated point. Then their product $X \times Y$ is countably tight.*

As indicated above, the assumption of Lindelöfness in (Q) is essential. We suggest the reader also deduce (Q) from the partition property (P) which will be considered in the next Chapter. This deduction is almost identical to the proof of Theorem 6 of [54] and it actually gives the stronger conclusion that the product

$$C_p(X) \times C_p(Y)$$

is Lindelöf. Here $C_p(X)$ and $C_p(Y)$ are the spaces of all continuous real functions on X and Y, respectively with the topology of pointwise convergence (see [75]).

9. (S) AND (L) ARE DIFFERENT

It is well-known that in (S) and (L) we can restrict ourselves to zero-dimensional spaces. To see this, note that we may restrict ourselves to Tychonoff spaces of size $\aleph_1$ which are therefore zero-dimensional since the continuum is bigger than $\aleph_1$ (see §§0-3). Now, for any zero-dimensional Hausdorff space X we can consider the algebra X^* of all clopen subsets of X. By identifying clopen subsets of X with their characteristic functions we can consider X^* as a subspace of $\{0,1\}^X$. Then X^* is the *dual space* of X. A typical basic open subset of X^* is determined by two disjoint finite subsets F and G of X as follows

$$[F;G] = \{U \in X^* : F \subseteq U \ \& \ U \cap G = \varnothing\}.$$

9.0. PROPOSITION. *X is hereditarily separable in each of its finite powers iff X^* is hereditarily Lindelöf in each of its finite powers.*

9.1. PROPOSITION. *X is hereditarily Lindelöf in each of its finite powers iff X^* is hereditarily separable in each of its finite powers.*

PROOF OF 9.0 AND 9.1. Suppose

$$\{x^\alpha : \alpha < \omega_1\} \subseteq X^n \text{ and } \{B^\alpha : \alpha < \omega_1\} \subseteq (X^*)^n$$

are given such that $x_i^\alpha \in B_i^\alpha$ for all $\alpha < \omega_1$ and $i < n$. Then for all $\alpha, \beta < \omega_1$ and $i < n$, we have

$$x_i^\alpha \notin B_i^\beta \text{ iff } B_i^\beta \notin [\{x_i^\alpha\}; \varnothing]$$

Hence $\{x^\alpha : \alpha < \omega_1\}$ is right (left)-separated in X^n iff $\{B^\alpha : \alpha < \omega_1\}$ is left-(right)-separated in $(X^*)^n$. This proves the converse implications of both 9.0 and 9.1.

Suppose now we are given $\{B^\alpha : \alpha < \omega_1\} \subseteq (X^*)^n$ and for each $\alpha < \omega_1$ and $i < n$ a basic open set $[F_i^\alpha; G_i^\alpha]$ containing B_i^α. By a Δ-system argument there is no loss of generality assuming F_i^α's and G_i^α's are disjoint. Moreover we may assume that for some k_i and ℓ_i ($i < n$), $|F_i^\alpha| = k_i$ and $|G_i^\alpha| = \ell_i$ for all $i < n$. Assuming X has a well-ordering let $\langle F_i^\alpha \rangle$ and $\langle G_i^\alpha \rangle$ denote the members of X^{k_i} and X^{ℓ_i} (resp.) which enumerate F_i^α and G_i^α (resp.) according to this ordering. Then for all $\alpha, \beta < \omega$ and $i < n$

$$\langle \langle F_i^\alpha \rangle, \langle G_i^\alpha \rangle \rangle \in (B_i^\beta)^{k_i} \times (X \setminus B_i^\beta)^{\ell_i} \text{ iff}$$

$$B_i^\beta \in [F_i^\alpha; G_i^\alpha].$$

Thus, $\{B^\alpha : \alpha < \omega_1\}$ is left (right)-separated in $(X^*)^n$ iff

$$\langle \langle F_0^\alpha \rangle, \dots, \langle F_{n-1}^\alpha \rangle, \langle G_0^\alpha \rangle, \dots, \langle G_{n-1}^\alpha \rangle \rangle$$

is right (left)-separated in $X^{\Sigma(k_i+\ell_i)}$. This proves the direct implications of both 9.0 and 9.1.

Let $(S)_w$ be the weakening of (S) by letting the hypothesis of (S) to be true for any finite power of the given regular space. Similarly, we let $(L)_w$ be the weakening of (L) by letting the hypothesis to be true for any finite power of the considered space. Then, as an immediate consequence of 9.0 and 9.1, we have the following well-known fact.

9.2. THEOREM. *$(S)_w$ and $(L)_w$ are equivalent statements.*

This explains why most of the known methods for proving results about (S) also give results about (L), and conversely. Namely, this is so because most of the known results about (S) and (L) are, in fact, results about $(S)_w$ and $(L)_w$. As a result of this experience quite frequently in the literature one finds a question asking whether the Conjectures (S) and (L) are, in fact, equivalent statements (see, e.g., [36], [41; Question (3)], [75; p. 91, Problem 14], ...). In this section we shall present a result showing that (S) and (L) are not equivalent. The result and its proof might appear a bit technical, but their understanding seems to be unavoidable if one wants to understand why the proof of (S) in [54] doesn't "dualize" and which kind of a problem must be resolved in any approach to (L).

9.3. THEOREM. *There is a model of Set Theory and Martin's Axiom where (S) is true but (L) is false.*

PROOF. It will be helpful it the reader is familiar with the papers [54] and [0], but we shall try to make the proof as self-contained as possible. For a set A, put $\mathcal{C}_A$ to be the set of all mappings σ from some finite subset of A into 2 and consider it as a poset under the reverse inclusion. Let $f : \omega_1 \times \omega_1 \to 2$ be a V-$\mathcal{C}_{\omega_1 \times \omega_1}$-generic function. For $\alpha < \omega_1$ define a subset U_α of ω_1 such that

$$\beta \in U_\alpha \text{ iff } \beta = \alpha \text{ or } \beta > \alpha \text{ and } f(\alpha, \beta) = 1.$$

Let $U_\alpha^1 = U_\alpha$ and $U_\alpha^0 = \omega_1 \setminus U_\alpha$. Let ϵ be the topology on ω_1 generated by

$$\{U_\alpha^i : \alpha < \omega_1 \ \& \ i < 2\}$$

So a typical basic open set of ϵ is of the form

$$[\sigma] = \bigcap\{U_\alpha^{\sigma(\alpha)} : \alpha \in \text{dom }(\sigma)\},$$

where σ is in C_{ω_1}. It is easily seen that ω_1, ϵ contradicts (L) but we shall need a stronger property of ω_1, ϵ. To introduce such a property of ϵ we need some definitions similar to those of [0].

9.4. Definitions.

(1) If M and N are two finite structures, then $M < N$ means that any *ordinal* mentioned in M is smaller than any ordinal mentioned in N.

(2) A *matrix of pairs* is a finite sequence $\langle F_0, \dots, F_{n-1}\rangle$ where each F_i is a strictly increasing finite sequence of pairs of countable ordinals.

(3) If F is a strictly increasing sequence of pairs, let

$$C(F) = \{\delta : \forall \langle \alpha, \beta\rangle \in F \ (\alpha < \delta \Leftrightarrow \beta < \delta)\}.$$

(4) A matrix of pairs $\langle F_0, \dots, F_{n-1}\rangle$ is *separated by a club* C in ω_1 iff for all $i < n, F_i$ is separated by

$$D_i = C \cap C(F_0) \cap \cdots \cap C(F_{i-1}),$$

i.e., between any two ordinals mentioned in F_i there is an element from D_i.

(5) A *matrix of neighborhoods* is a finite sequence $\langle \Sigma_0, \dots, \Sigma_{m-1}\rangle$ where each Σ_i is a finite subset of C_{ω_1} such that $\bigcup \Sigma_i$ is a function, i.e., every two members of Σ_i are compatible in C_{ω_1}.

9.5. LEMMA. *The space ω_1, ϵ is 2-complicated in the following sense: There is a club C in ω_1 such that for any increasing sequence*

$$\langle M^\alpha = \langle F_0^\alpha, \dots, F_{n-1}^\alpha\rangle : \alpha < \omega_1\rangle$$

of matrices of pairs separated by C and any increasing sequence

$$\langle \Sigma^\alpha = \langle \Sigma_0^\alpha, \dots, \Sigma_{n-1}^\alpha\rangle : \alpha < \omega_1\rangle$$

of matrices of neighborhoods there exist $\alpha < \beta$ such that

$$\Phi(\Sigma^\alpha; M^\beta) : \forall i < n \ \forall \sigma \in \Sigma_i^\alpha \ \forall \langle x, y\rangle \in F_i^\beta \ [\sigma] \cap \{x, y\} \neq \varnothing.$$

PROOF. The proof is very similar to the corresponding proof in [0]. It is done in V via a forcing argument for the poset $C_{\omega_1 \times \omega_1}$. So fix some names $\{\dot{M}^\alpha : \alpha < \omega_1\}$ and $\{\dot{\Sigma}^\alpha : \alpha < \omega_1\}$ and a condition p in $C_{\omega_1 \times \omega_1}$ forcing that they satisfy the hypothesis of 9.5. For each α, find $p_\alpha \leq p$ and matrices M^α and Σ^α such that p_α forces

$$\dot{M}^\alpha = M^\alpha \text{ and } \dot{\Sigma}^\alpha = \Sigma^\alpha.$$

We may assume p_α contains all ordinals mentioned in M^α and Σ^α. Moreover we may assume that p_α's form a Δ-system. Note that we may assume that no ordinal mentioned in M^α or Σ^α, $(\alpha < \omega_1)$ is in the root. The main reason for the definition of a separated matrix of pairs is that for each $\alpha < \omega_1$ we can find a sequence S_i^α $(i < n)$ of disjoint finite sets such that

$$\forall\, i < n \,\forall\, \langle x, y \rangle \in F_i^\alpha \ S_i^\alpha \cap \{x, y\} \neq \emptyset.$$

The existence of S_i^α's follows from Lemma 5.3 of [0] and it is proven in an easy induction on the size of the matrix. It is clear now that given any $\alpha < \beta$ we can find $q \leq p_\alpha, p_\beta$ such that

$$\forall\, i < n \,\forall\, \sigma \in \Sigma_i^\alpha \,\forall\, x \in \mathrm{dom}(\sigma) \,\forall\, y \in S_i^\beta \ q(x, y) = \sigma(x).$$

The fact that every two members of Σ_i^α are compatible is used here. It is clear that q forces $\Phi(\dot\Sigma^\alpha; \dot M^\beta)$. This completes the proof.

From now on, we assume that our ground model V satisfies GCH and has the above 2-complicated space ω_1, ϵ. Let $\langle P_\alpha : \alpha \leq \omega_2 \rangle$ be the mixed iteration of the poset C_{ω_1} at even stages with finite supports and the Jensen club-set poset $\mathcal{E}$ at odd stages with countable supports (see [54]) and let $P = P_{\omega_2}$.

9.6. LEMMA. *The space* ω_1, ϵ *remains 2-complicated in* V^P.

PROOF. We know that P is regularly embeddable into a product of the form $C_A \times \tilde{P}$, where $A = \omega_2 \times \omega_1$ and where $\tilde{P}$ is a σ-closed poset. So, the standard Easton-type argument (see [54]) reduces the problem to the problem of preserving 2-complicatedness of ω_1, ϵ in V^{C_A}. This follows easily from the fact that C_A is a property K poset.

The model which will satisfy the conclusion of 9.3 will be a finite support iteration in V^P of certain "nondangerous" ([0]) *ccc* posets of size $\aleph_1$. The main difficulty is to handle the following partition property which easily implies (S) (see [54] or the end of this chapter) with a nondangerous *ccc* poset.

(P) *For any partition* $[\omega_1]^2 = K_0 \cup K_1$ *either there is an uncountable* $A \subseteq \omega_1$ *such that* $[A]^2 \subseteq K_0$, *or else there exist uncountable* $B \subseteq \omega_1$ *and disjoint uncountable* $\mathcal{F} \subseteq [\omega_1]^{<\omega}$ *such that for all* α *in* B *and* F *in* $\mathcal{F}$ *with* $\alpha < F$ *there exists a* β *in* F *such that* $\{\alpha, \beta\} \in K_1$.

(The pairs $\langle B, \mathcal{F} \rangle$ with the above property will be called *bad pairs*.) The following lemma shows how to do this and will take the role of Lemma 1 of [54] in finishing the proof of 9.3.

9.7. LEMMA. *Let* $C = C_{\omega_1}$ *and let* G_C *be a V-generic subset of* C. *In* $V[G_C]$, *let* $\mathcal{E}$ *be the club-set-poset. Let* Q *be a ccc poset such that* ω_1, ϵ *is 2-complicated in* $V[G_Q]$, *where* G_Q *is a* $V[G_C]$-*generic subset of* Q. *Let*

(1) $[\omega_1]^2 = K_0 \cup K_1$

be a partition in $V[G_{\mathcal{Q}}]$. Let $G_{\mathcal{E}}$ be a $V[G_C][G_{\mathcal{Q}}]$-generic subset of $\mathcal{E}$. Let $C_{\mathcal{E}}$ be the corresponding club in $V[G_C][G_{\mathcal{Q}}][G_{\mathcal{E}}]$ where we define

$$S = \{s \in [\omega_1]^{<\omega} : [s]^2 \subseteq K_0 \text{ and } s \text{ is separated by } C_{\mathcal{E}}\}$$

with the ordering $\supseteq$. Suppose either that S is not a ccc poset, or else that ω_1, ϵ is not 2-complicated in $V[G_C][G_{\mathcal{Q}}][G_{\mathcal{E}}][G_S]$. Then in $V[G_C][G_{\mathcal{Q}}]$, we can find a ccc poset R which forces a bad pair for the given partition (1) such that ω_1, ϵ is still a 2-complicated space in $V[G_C][G_{\mathcal{Q}}][G_R]$.

PROOF. Suppose that the poset S is either not *ccc* or else that ω_1, ϵ is not 2-complicated in

$$V[G_C][G_{\mathcal{Q}}][G_{\mathcal{E}}][G_S]$$

with respect to the club $E \cap C_{\mathcal{E}}$, where $E \in V$ is a club which shows ω_1, ϵ is 2-complicated in $V[G_C][G_{\mathcal{Q}}][G_{\mathcal{E}}]$. This means that, working in $V[G_C][G_{\mathcal{Q}}][G_{\mathcal{E}}]$, we can find a sequence $\{s_\alpha : \alpha < \omega_1\}$ of members of S and strictly increasing sequences $\{M^\alpha : \alpha < \omega_1\}$ and $\{\Sigma^\alpha : \alpha < \omega_1\}$ of matrices of pairs separated by $E \cap C_{\mathcal{E}}$ and matrices of neighborhoods, respectively, such that all M^α's and Σ^α's have m columns for some fixed $m \geq 1$ and such that for all $\alpha < \beta < \omega_1$

(a) $\Phi(\Sigma^\alpha; M^\beta)$ implies $s_\alpha \perp s_\beta$ in S.

We may assume that s_α's form a Δ system with root s and such that if $t_\alpha = s_\alpha \setminus s$, then t_α's are strictly increasing and of the same cardinality $n \geq 1$.

From now on we shall work in $V[G_C][G_{\mathcal{Q}}]$ fixing $\mathcal{E}$-names $\{\dot{s}_\alpha : \alpha < \omega_1\}$, $\{\dot{t}_\alpha : \alpha < \omega_1\}$, $\{\dot{M}^\alpha : \alpha < \omega_1\}$ and $\{\dot{\Sigma}^\alpha : \alpha < \omega_1\}$ for the above objects and a condition $\langle a_0, A_0 \rangle$ in $\mathcal{E}$ forcing all of the above facts about them.

Let θ be a large enough regular cardinal and let N_0 be a countable elementary submodel of $\mathbf{H}_\theta$ containing everything relevant, such that

$$N_0 \cap V[G_C] \in V[G_C].$$

Let $\delta_0 = N_0 \cap \omega_1$, let $F \in [\omega_1 \setminus \delta_0]^n$, and let

$$\bar{M} = \langle \bar{F}_0, \dots, \bar{F}_{m-1} \rangle$$

be a matrix of pairs separated by E having all ordinals above δ_0. Let $\mathcal{W}_{F,\bar{M}}$ be the set of all $\langle a, A \rangle$ in $\mathcal{E} \cap N_0$ such that either $\langle a, A \rangle \perp \langle a_0, A_0 \rangle$, or else $\langle a, A \rangle \leq \langle a_0, A_0 \rangle$ and there exist $\xi < \delta_0$ and t and Σ in N_0 such that

(b) $t \times F \subseteq K_0$, $\Phi(\Sigma; \bar{M})$ and $\langle a, A \rangle \Vdash \dot{t}_\xi = t \,\&\, \dot{\Sigma}_\xi = \Sigma$

CLAIM. *There exist F and $\bar{M}$ for which $\mathcal{W}_{F,\bar{M}}$ is not dense in $\mathcal{E} \cap N_0$.*

PROOF. Suppose $\mathcal{W}_{F,\bar{M}}$ is dense for all F and $\bar{M}$. Define $R \subseteq (N_0)^4$ by

$$\langle \langle a, A \rangle, \xi, t, \Sigma \rangle \in R \text{ iff } \langle a, A \rangle \Vdash \dot{t}_\xi = t \,\&\, \dot{\Sigma}_\xi = \Sigma.$$

Since $\mathcal{E} \cap N_0$ and R can be coded using only a countable amount of information there is an $\alpha < \omega_1$ such that $\mathcal{E} \cap N_0$ is in $V[G_{C_\alpha}]$ and R is in $V[G_{C_\alpha}][G_{\mathcal{Q}}]$. Note that this means that for all F and $\bar{M}$, the set $\mathcal{W}_{F,\bar{M}}$ is in $V[G_{C_\alpha}][G_{\mathcal{Q}}]$

since it is definable from $F, \bar{M}, K_0, K_1, \mathcal{E} \cap N_0, \langle a_0, A_0 \rangle, \delta_0$, and R. The poset $\mathcal{E} \cap N_0$ is countable therefore we can find a mapping π in $\mathbf{V}[G_{C_n}]$ which densely embeds $C_{[\alpha, \alpha + \omega)}$ into the set of all elements of $\mathcal{E} \cap N_0$ below $\langle a_0, A_0 \rangle$. The restriction $G_{\alpha, \alpha + \omega}$ of G_C to $C_{[\alpha, \alpha + \omega)}$ is $\mathbf{V}[G_{C_n}][G_Q]$-generic, so $\pi'' G_{\alpha, \alpha + \omega}$ is a directed subset of $\mathcal{E} \cap N_0$ which intersect all $W_{F, \bar{M}}$'s. Since it is countable and since it is an element of $\mathbf{V}[G_C]$, it has a lower bound $\langle \bar{a}, \bar{A} \rangle$ in $\mathcal{E}$. Pick $\langle b, B \rangle \leq \langle \bar{a}, \bar{A} \rangle$ and $F, \bar{M}$ above δ such that for some η,

$$\langle b, B \rangle \Vdash \dot{i}_\eta = F \ \& \ \dot{\bar{M}}^\eta = \bar{M}.$$

By the choice of $\langle \bar{a}, \bar{A} \rangle$ there is an $\langle a, A \rangle$ in $W_{F, \bar{M}}$ such that $\langle b, B \rangle \leq \langle a, A \rangle$. By (b), we can find $\xi < \delta_0$ and t, Σ in N_0 such that

$$t \times F \subseteq K_0, \ \Phi(\Sigma; \bar{M}), \ \text{and}$$

$$\langle a, A \rangle \Vdash \dot{i}_\xi = t \ \& \ \dot{\Sigma}_\xi = \Sigma.$$

This means that $\langle b, B \rangle$ forces

$$\Phi(\dot{\Sigma}_\xi; \dot{\bar{M}}^\eta) \ \& \ \dot{s}_\xi \cup \dot{s}_\eta \in \dot{S}$$

contradicting (a). This proves the Claim.

So fix such F and $\bar{M}$ and let $\langle b, B \rangle \leq \langle a_0, A_0 \rangle$ be a condition in $\mathcal{E} \cap N_0$ having no extension in $W_{F, \bar{M}}$. Let T be the subtree of $[\omega_1]^{\leq n}$ (with the ordering $\leq \cdot$ defined by: $s \leq \cdot t$ iff s is an initial part of t) such that for every t in $T \cap [\omega_1]^n$ there exist

$$\langle a(t), A(t) \rangle \leq \langle b, B \rangle, \ \xi = \xi(t) < \omega_1, \ \text{and} \ \Sigma = \Sigma(t)$$

such that

$$\langle a(t), A(t) \rangle \Vdash \dot{i}_\xi = t \ \& \ \dot{\Sigma}^\xi = \Sigma.$$

Note that T is a member of N_0 being definable from parameters in N_0. Since the conditions of $\dot{S}$ are separated by the generic club $C_\mathcal{E}$ which is almost included in any club from $\mathbf{V}[G_C]$ (and, therefore, almost included in any club from $\mathbf{V}[G_C][G_Q]$), the following is true about our tree T:

(c) For every club D in ω_1 there is a t in $T \cap [\omega_1]^n$ such that D separates the points of t.

It is easily seen that (c) implies that $T \cap [\omega_1]^n$ with the lexicographical ordering has type ω_1^n. So as before, by shrinking T, we may assume that T is everywhere $\aleph_1$ splitting. Note that

(d) $\forall t \in T \cap [\omega_1]^n \cap N_0 [\Phi(\Sigma(t); \bar{M}) \Rightarrow t \times F \not\subseteq K_0]$.

Thus, using the elementarity of N_0 and a further refinement of the tree T, we may assume to have strictly increasing sequence

$$\{\bar{M}^\alpha = \langle \bar{F}_0^\alpha, \ldots, \bar{F}_{m-1}^\alpha \rangle : \alpha < \omega_1\}$$

of matrices of pairs separated by E and a strictly increasing sequence $\{F_\alpha : \alpha < \omega_1\}$ of elements of $[\omega_1]^n$ such that

(e) $\forall t \in T \cap [\omega_1]^n \ \forall \alpha < \omega_1 [t < \bar{M}^\alpha \ \& \ \Phi(\Sigma(t); \bar{M}^\alpha) \Rightarrow t \times F_\alpha \not\subseteq K_0]$.

Furthermore, by going to a subtree of T, we may assume that the matrices

$$\Sigma(t) = \langle \Sigma_0(t), \dots, \Sigma_{m-1}(t) \rangle, \ t \in T \cap [\omega_1]^n$$

satisfy

(f) $\forall s, t \in T \cap [\omega_1]^n \ \forall \ i < m [\bigcup \Sigma_i(s)$ and $\bigcup \Sigma_i(t)$ are compatible in $C_{\omega_1}]$.

Note that

$$D = E \cap \bigcap \{ C(\bar{F}_0^\alpha) \cap \cdots \cap C(\bar{F}_{m-1}^\alpha) : \alpha < \omega_1 \}$$

is a club in ω_1 since a given α can effect E only in a fixed interval and the corresponding intervals for distinct α are disjoint.

Let $\mathcal{R}$ be the set of all pairs $\langle X, Y \rangle$ such that:

(g) X is a finite subset of $T \cap [\omega_1]^n$,
(h) Y is a finite subset of ω_1,
(i) $\forall \ t \in X \ \forall \ \alpha \in Y [t < \bar{M}^\alpha \Rightarrow \Phi(\Sigma(t); \bar{M}^\alpha)]$.

We order $\mathcal{R}$ by letting $\langle X_0, Y_0 \rangle \leq \langle X_1, Y_1 \rangle$ iff $X_0 \supseteq X_1$ and $Y_0 \supseteq Y_1$.

CLAIM. $\mathcal{R}$ *is a ccc poset,* ω_1, ϵ *is 2-complicated after forcing by* $\mathcal{R}$, *and the club* D *shows this.*

PROOF. Suppose we have a sequence $\{ \langle X_\xi, Y_\xi \rangle : \xi < \omega_1 \}$ of elements of $\mathcal{R}$, a strictly increasing sequence

$$\{ G^\xi = \langle G_0^\xi, \dots, G_{p-1}^\xi \rangle : \xi < \omega_1 \}$$

of matrices of pairs separated by D and a strictly increasing sequence

$$\{ \Omega^\xi = \langle \Omega_0^\xi, \dots, \Omega_{p-1}^\xi \rangle : \xi < \omega_1 \}$$

of matrices of neighborhoods. We have to find ξ and ζ such that $\langle X_\xi, Y_\xi \rangle$ and $\langle X_\zeta, Y_\zeta \rangle$ are compatible in $\mathcal{R}$ and such that $\Phi(\Omega^\xi; G^\zeta)$ holds. Note that this will prove both parts of the Claim. By a standard Δ system argument we may assume that for all $\xi < \zeta$, any ordinal appearing in X_ξ, Y_ξ, G^ξ or Ω^ξ is less than any ordinal appearing in $X_\zeta, Y_\zeta, G^\zeta$ or Ω^ζ.

For $\xi < \omega_1$, we set

$$M^{*\xi} = \langle \bigcup_{\alpha \in Y_\xi} F_0^\alpha, \dots, \bigcup_{\alpha \in Y_\xi} F_{m-1}^\alpha, G_0^\xi, \dots, G_{p-1}^\xi \rangle, \text{ and}$$

$$\Sigma^{*\xi} = \langle \bigcup_{t \in X_\xi} \Sigma_0(t), \dots, \bigcup_{t \in X_\xi} \Sigma_{m-1}(t), \Omega_0^\xi, \dots, \Omega_{p-1}^\xi \rangle.$$

Note that by the assumption about G^ξ and our choice of D, the matrix $M^{*\xi}$ is separated by E. Note also that $\Sigma^{*\xi}$ is a matrix of neighborhoods by (f). Since ω_1, ϵ is 2-complicated with respect to the club E, we can find $\xi < \zeta$ such that

(j) $\Phi(\Sigma^{*\xi}; M^{*\xi})$ holds.

It is now easily checked that (j) implies:

(k) $\langle X_\xi \cup X_\zeta, Y_\xi \cup Y_\zeta \rangle$ is a condition of $\mathcal{R}$, and
(ℓ) $\Phi(\Omega^\xi; G^\zeta)$ holds.

This finishes the proof of the Claim.

Let G_R be a $V[G_C][G_Q]$-generic subset of R and let in $V[G_C][G_Q][G_R]$

$$T^* = \{s \in [\omega_1]^{\leq n} : \exists \langle X, Y \rangle \in G_R \; \exists t \in X \; s \leq \cdot t\}, \text{ and}$$

$$Y^* = \bigcup \{Y : \exists X \langle X, Y \rangle \in G_R\}.$$

CLAIM. *There is an $\langle X_0, Y_0 \rangle$ in R which forces that $\dot{T}^*$ contains an everywhere $\aleph_1$-splitting subtree of height $n + 1$ and that $\dot{Y}^*$ is uncountable.*

PROOF. Otherwise, since R is a ccc poset, the tree T would be the union of countably many trees none of which contain an everywhere $\aleph_1$-splitting subtree of height $n + 1$. But this is clearly impossible.

Thus we shall actually force with the poset

$$R\langle X_0, Y_0 \rangle = \{\langle X, Y \rangle \in R : \langle X, Y \rangle \leq \langle X_0, Y_0 \rangle\}$$

rather than R. But we shall use the letter R to denote also $R\langle X_0, Y_0 \rangle$ and we shall use T^* to denote an everywhere $\aleph_1$-splitting subtree of T^* of height $n + 1$. Note that by (e) and (i) in $V[G_C][G_Q][G_R]$, we have

(m) $\forall t \in T^* \cap [\omega_1]^n \; \forall \alpha \in Y^* \; (t < \bar{M}^\alpha \Rightarrow t \times F_\alpha \not\subseteq K_0)$.

Let

$$A_\varnothing = \{\xi < \omega_1 : \{\xi\} \in T^*\}$$

Then $A_\varnothing$ is uncountable. If there is a ξ_0 in A_0 and uncountable $Z_0 \subseteq Y^*$ such that

$$\{\xi_0\} \times F_\beta \subseteq K_0 \text{ for all } \beta \text{ in } Z_0,$$

we consider

$$A_{\langle \xi_0 \rangle} = \{\xi < \omega_1 : \{\xi_0, \xi\} \in T^*\}.$$

Then $A_{\langle \xi_0 \rangle}$ is also uncountable and we ask for a ξ_1 in $A_{\langle \xi_0 \rangle}$ and uncountable $Z_1 \subseteq Z_0$ such that

$$\{\xi_1\} \times F_\beta \subseteq K_0 \text{ for all } \beta \text{ in } Z_1,$$

and so on. This process must terminate before we reach a top node $t = \{\xi_0, \dots, \xi_{n-1}\}$ of T^* or else we get a contradiction with (m). This means that we can find uncountable $A \subseteq \omega_1$ and $Z \subseteq Y^*$ such that

(n) $\forall \alpha \in A \; \forall \beta \in Z \; [\alpha < F_\beta \Rightarrow \exists \gamma \in F_\beta \; \{\alpha, \gamma\} \in K_1]$.

This means that in $V[G_C][G_Q][G_R]$ we have a bad pair for the partition (1). Hence the poset R satisfies the conclusion of Lemma 9.7, and the proof is finished.

We are now ready to finish the proof of 9.3. Working in V^p, we build a finite support ω_2-iteration of ccc nondangerous posets of size $\aleph_1$, i.e., posets which preserve 2-complicatedness of the space ω_1, ϵ. Note that 9.7 is just what is needed in showing that such an iteration can indeed force (P) (and, therefore, (S)) to hold in the final model. To take care about MA, all we have to show is that we can kill the *ccc* property of a dangerous poset $\mathcal{Q}$ with a nondangerous one. So, suppose our space ω_1, ϵ is 2-complicated with respect to a club C, but it fails to be 2-complicated with respect to C in the forcing extension by $\mathcal{Q}$. So, we can find a sequence $\{q_\alpha : \alpha < \omega_1\}$ of elements of $\mathcal{Q}$, a strictly increasing sequence of matrices $\{M^\alpha : \alpha < \omega_1\}$ of pairs separated by C, and a strictly increasing sequence $\{\Sigma^\alpha : \alpha < \omega_1\}$ of matrices of neighborhoods (with M^α and Σ^α having the same number of columns) such that for all $\alpha < \beta < \omega_1$,

(o) $\Phi(\Sigma^\alpha; M^\beta)$ implies $q_\alpha \perp q_\beta$.

Let $\mathcal{R}$ be the set of all finite $r \subseteq \omega_1$ such that $\Phi(\Sigma^\alpha; M^\beta)$ holds for all $\alpha < \beta$ in r. The order on $\mathcal{R}$ is $\supseteq$. Then an argument similar to that for the poset $\mathcal{R}$ in the proof of 9.7 shows that $\mathcal{R}$ is *ccc* and nondangerous. Note that by (o), $\mathcal{R}$ kills the *ccc* property of the poset $\mathcal{Q}$.

We can arrange our space ω_1, ϵ to have the following combinatorial property in the final model:

(∗) Suppose we are given uncountable $X \subseteq \omega_1$ and a σ mapping X into $\mathcal{C}_{\omega_1}$ such that $\operatorname{dom}\sigma(x) < \operatorname{dom}\sigma(y)$ whenever $x < y$. Then for every $f : X \times \omega_1 \to \omega_1$ with property $f(x, \alpha) > \alpha$ for all x and α, there is a strictly increasing sequence $\{\langle x_\alpha, \xi_\alpha, y_\alpha \rangle : \alpha < \omega_1\}$ of elements of $X \times \omega_1 \times \omega_1$ such that for all $\alpha < \beta < \omega_1$:

(p) $x_\alpha < \xi_\alpha < y_\alpha$,

(q) $f(x_\alpha, \xi_\alpha) = y_\alpha$,

(r) $[\sigma(x_\alpha)] \cap \{x_\beta, y_\beta\} \neq \varnothing$.

To see that this can be arranged, suppose that we are at some intermediate stage of the *ccc* iteration and that we are given X, σ and f. Let C be a club witnessing the fact that ω_1, ϵ is 2-complicated. Pick an increasing sequence $\{\langle x_\alpha, \xi_\alpha, y_\alpha \rangle : \alpha < \omega_1\}$ of elements of $X \times \omega_1 \times \omega_1$ such that ξ_α's are in C and (p) and (q) hold. Moreover, we assume that for all $\alpha < \beta$ there is an element of C separating the two sets

$$\operatorname{dom}\sigma(x_\alpha) \cup \{x_\alpha, \xi_\alpha, y_\alpha\} \text{ and } \operatorname{dom}\sigma(x_\beta) \cup \{x_\beta, \xi_\beta, y_\beta\}.$$

Let $\mathcal{R}$ be the poset of all finite $r \subseteq \omega_1$ such that for all $\alpha < \beta$ in r the condition (r) holds. Then $\mathcal{R}$ is *ccc* and the space ω_1, ϵ is 2-complicated in the forcing extension of $\mathcal{R}$. The proofs of these two facts are again almost identical to the proofs of the corresponding facts for the poset $\mathcal{R}$ from 9.7. Note that, if we force with $\mathcal{R}$ at this stage, we get the conclusion of (∗) for the X, σ and f we have started with.

9.8. PROPOSITION. *If (∗) is true about ω_1, ϵ, then for every sequence $\{\sigma_\alpha : \alpha < \omega_1\} \subseteq \mathcal{C}_{\omega_1}$ with property $\operatorname{dom}\sigma_\alpha < \operatorname{dom}\sigma_\beta$ for $\alpha < \beta$, there is a γ such that every $\beta \geq \gamma$ is in some $[\sigma_\alpha]$ for $\alpha < \gamma$.*

PROOF. Otherwise, for every γ we can pick $x_\gamma \geq \gamma$ not in any $[\sigma_\alpha]$ for $\alpha < \gamma$. Let X be equal to $\{x_\gamma : \gamma < \omega_1\}$ and let $\sigma : X \to \mathcal{C}_{\omega_1}$ be defined by $\sigma(\alpha) = \sigma_\alpha$. For $\langle x, \alpha \rangle$ in $X \times \omega_1$ let $f(x, \alpha)$ be the minimal member of X above α. By $(*)$, we have a sequence

$$\{\langle x(\alpha), \xi(\alpha), y(\alpha) \rangle : \alpha < \omega_1\}$$

satisfying (p), (q) and (r). Pick $\alpha < \beta < \omega_1$ with the property that if $x(\beta) = x_\gamma$ and $y(\beta) = x_\delta$, then $x(\alpha) < \gamma, \delta$. By (r) and the way σ is defined we have that one of the x_γ or x_δ is in $[\sigma_{x(\alpha)}]$, a contradiction.

Note that 9.8 shows that our space ω_1, ϵ in the final model is hereditarily Lindelöf (and, therefore, a counterexample to (L)) in a very strong sense.

9.9. PROPOSITION. *If ω_1, ϵ satisfies $(*)$, then $(*)$ is true about ω_1, ϵ in any ccc forcing extension.*

PROOF. Let $\mathcal{Q}$ be a ccc poset and fix a $\mathcal{Q}$-names $\dot{X}$, $\dot{\sigma}$ and $\dot{f}$ which are forced to satisfy the hypothesis of $(*)$. Find uncountable $A \subseteq \omega_1$, and increasing $\rho : A \to \mathcal{C}_{\omega_1}$ and for each a in A a condition p_a in $\mathcal{Q}$ such that

$$p_a \Vdash a \in \dot{X} \ \& \ \dot{\sigma}(a) = \rho(a).$$

Define $g : A \times \omega_1 \to \omega_1$, by letting $g(a, \xi)$ be the minimal member b of ω_1 such that some $p(a, \xi, b) \leq p_a$ forces $\dot{f}(a, \xi) = b$. By $(*)$ we can find increasing

$$\{\langle a_\alpha, \xi_\alpha, b_\alpha \rangle : \alpha < \omega_1\}$$

satisfying (p), (q) and (r) for A, ρ and g in place of X, σ and f. Since $\mathcal{Q}$ is ccc there is a p in $\mathcal{Q}$ which forces

$$\dot{E} = \{\alpha < \omega_1 : p(a_\alpha, \xi_\alpha, b_\alpha) \in \dot{G}\}$$

is uncountable. Thus, p forces that

$$\{\langle a_\alpha, \xi_\alpha, b_\alpha \rangle : \alpha \in \dot{E}\}$$

is an increasing sequence of triples satisfying the conclusion of $(*)$ for $\dot{X}, \dot{\sigma}$ and $\dot{f}$. Since we could have started with any condition in $\mathcal{Q}$ this completes the proof.

Note that the above two facts 9.8 and 9.9 show that the property $(*)$ of the space ω_1, ϵ contradicting (L) behaves in a similar way to the property $(*)$ of the first countable counterexample to (S) constructed in [0] in a model of $MA_{\aleph_1}$.

The partition property (P) has the conclusion of 8.10 as an immediate consequence, so we have that the conclusions of 8.11 and 8.12 are both true in the model of 9.3. To see this, suppose X is an arbitrary topological space which has an open cover $\mathcal{U}$ such that

$$X \neq \bigcup \{\bar{U} : U \in \mathcal{U}_0\}$$

for every countable $\mathcal{U}_0 \subseteq \mathcal{U}$. So, recursively on $\alpha < \omega_1$ we can pick sequences $\{U_\alpha : \alpha < \omega_1\}$ and $\{x_\alpha : \alpha < \omega_1\}$ of members of $\mathcal{U}$ and X, respectively such that:

(a) $x_\alpha \in U_\alpha$,

(b) $x_\alpha \notin \bigcup_{\xi < \alpha} \bar{U}_\xi$.

Define $[\omega_1]^2 = K_0 \cup K_1$ by

$$\{\alpha, \beta\}_< \in K_0 \text{ iff } x_\alpha \notin U_\beta.$$

Applying (P) to this partition gives us an uncountable discrete subspace of X.

9.10. Remarks.

The fact that $(S)_w$ and $(L)_w$ are equivalent statements is a result of Zenor [67] and Roitman [36], where a slightly different notion of a dual space is considered. We suggest the reader use the duality to explain the symmetry in 7.10 and 7.13. Note that 7.11 shows that $(S)_w$ and $(L)_w$ are consequences of $\text{MA}_{\aleph_1}$.

The main result 9.3 of this section first appeared in our note [62g].

The proof of Theorem 9.3 clearly shows the great advantage of using the mixed iteration rather than a countable support iteration in the proof of (S) in [54]. The experience shows that, even if we don't have general iteration theorems, the main difficulty in proving results of this kind usually lies in designing the single-step poset.

REFERENCES

[0] U. Abraham and S. Todorcevic, Martin's Axiom and first countable S and L spaces, in *Handbook of Set-Theoretic Topology* (K. Kunen and J. Vaughan, eds.), Elsevier Science Publishers B.V., (1984), 327-346.

[1] U. Abraham, M. Rubin and S. Shelah, On the consistency of some partition theorems for continuous colorings, and the structure of $\aleph_1$-dense real order types, *Ann. Pure Appl. Logic* 29 (1985), 123-206.

[2] J. E. Baumgartner, All $\aleph_1$-dense sets of reals can be isomorphic, *Fund. Math.* 79 (1973), 101-106.

[3] J. E. Baumgartner, Chain and antichains in $P(\omega)$, *J. Symbolic Logic* 45 (1980), 85-92.

[4] J. E. Baumgartner, Applications of the Proper Forcing Axiom, in *Handbook of Set Theoretic Topology* (K. Kunen and J. Vaughan, eds.), Elsevier Science Publishers B.V. (1984), 913-959.

[5] K. J. Devlin, The Yorkshireman's guide to proper forcing, in *Surveys in Set Theory* (A. R. D Mathias, ed.), London Math. Soc. Lect. Note Series: 87, Cambridge Univ. Press, London 1983.

[6] E. K. van Douwen, F. Tall and W. Weiss, Nonmetrizable hereditarily Lindelöf spaces with point countable bases from CH, *Proc. Amer. Math. Soc.* 64 (1977), 139-145.

[7] E. K. van Douwen and K. Kunen, L-spaces and S-spaces in $P(\omega)$, *Topology Appl.* 14 (1982), 143-149.

[8] E. K. van Douwen, The Integers and Topology, in *Handbook of Set-Theoretic Topology* (K. Kunen and J. Vaughan, eds.), Elsevier Science Publishers B.V., (1984), 111-167.

[9] A. Dow, PFA and ω_1^*, *Topology Appl.* 28 (1988), 127-140.

[10] R. Engelking, *General Topology*, PWN, Warszawa 1977.

[11] P. Erdös and A. Tarski, On the families of mutually exclusive sets, *Ann. of Math.* 44 (1943), 315-329.

[12] D. Fremlin, *Consequences of Martin's Axiom*, Cambridge Univ. Press 1984.

[13] D. Fremlin, Applications of Martin's maximum, preprint 1986.

[14] V. Fedorchuk, On the cardinality of hereditarily separable compact Hausdorff spaces, *Soviet Math. Doklady* 16 (1975), 651-655.

[15] R. J. Gardner and W. F. Pfeffer, Are diffused, regular, Radon measures σ-finite?, *J. London Math. Soc.* 20(2) (1979), 485-494.

[16] R. J. Gardner and W. F. Pfeffer. Borel Measures, in *Handbook of Set-Theoretic Topology* (K. Kunen and J. Vaughan, eds.), Elsevier Science Publishers B.V., (1984), 961-1043.

[17] J. de Groot, Discrete Subspaces of Hausdorff Spaces, *Bull. Acad. Polon. Sci.* 13 (1965), 537-544.

[18] G. Gruenhage, Cosmicity of Cometrizable spaces, preprint 1987.

[19] A. Hajnal and I. Juhász, On discrete subspaces of topological spaces, *Indag. Math.* 29 (1967), 343-356.

[20] F. Hausdorff, Summen von $\aleph_1$ Mengen, *Fund. Math.* 26 (1936), 241-255.

[21] I. Juhász, A survey of S and L spaces, *Coll. Math. Soc. János Bolyai* 23, Topology, Budapest 1978, North-Holland, Amsterdam (1980), 674-688.

[22] I. Juhász, Cardinal Functions II, in *Handbook of Set-Theoretic Topology* (K. Kunen and J. Vaughan, eds.), Elsevier Science Publishers B.V. (1984), 63-109.

[23] I. Juhász, K. Kunen and M. E. Rudin, Two more hereditarily separable non-Lindelöf spaces, *Can. J. Math.* 28 (1976), 998-1005.

[24] K. Kunen, Products of S-spaces, preprint May 1975.

[25] K. Kunen, (κ, λ^*) gaps under MA, preprint August 1976.

[26] K. Kunen, Luzin spaces, *Topology Proc.* 1 (1976), 191-199.

[27] K. Kuratowski, *Topology*, Vol. 1 (Academic Press, New York), 1966.

[28] K. Kuratowski and W. Sierpinski, Le théorème de Borel-Lebesgue dans la théorie des ensembles abstraits, *Fund. Math.* 2 (1921), 172-178.

[29] D. Kurepa, Ensembles ordonnés et ramifiés, *Publ. Inst. Math. Univ. Belgrade* 4 (1935), 1-138.

[30] D. Kurepa, Dendrity of Spaces and of Ordered Sets, *Glasnik Matematicki* Ser. III, 2(22) (1967), 145-162.

[31] D. Kurepa, Around the general Souslin problem, *Proceedings of the International Symposium on Topology in Herceg-Novi 1968*, Beograd (1969), 239-245.

[32] E. Michael, Paracompactness and the Lindelöf property in finite and countable Cartesian products, *Compositio Math.* 23 (1971), 119-214.

[33] A. Ostaszewski, On countably compact, perfectly normal spaces, *J. London Math. Soc.* (2)14 (1976), 505-516.

[34] J. Roitman, A reformulation of S and L, *Proc. Amer. Math. Soc.* 69 (1970).

[35] J. Roitman, The spread of regular spaces, *General Topology and Appl.* 8 (1978), 85-91.

[36] J. Roitman, Adding a random or a Cohen real, *Fund. Math.* 103 (1979), 47-60.

[37] J. Roitman, Basic S and L, in *Handbook of Set-Theoretic Topology* (K. Kunen and J. Vaughan, eds.), Elsevier Science Publishers B.V. (1984), 295-326.

[38] F. Rothberger, Sur les famillies indënombrables de suites de nombres naturels et les problèmes concernant la propriété C, *Proc. Cambridge Phil. Soc.* 37 (1941), 109-126.

[39] M. E. Rudin, A normal hereditarily separable non-Lindelöf spaces, *Illinois J. Math.* 16 (1972), 621-626.

[40] M. E. Rudin, Lectures on set theoretic topology, *CBMS Regional Conf. Ser.* 23, (Amer. Math. Soc., Providence) 1975.

[41] M. E. Rudin, S and L spaces, in *Set-Theoretic Topology* (G.M. Reed, ed.), Academic Press, New York (1980), 431-444.

[42] L. Schwartz, *Radon Measures on Arbitrary Topological Spaces and Cylindrical Measures*, Oxford Univ. Press, London 1973.

[43] S. Shelah, Decomposing uncountable squares to countably many chains, *J. Comb. Theory* (A) 21(1976), 110-114.

[44] S. Shelah, Proper Forcing, *Lecture Notes in Math.* 940, Springer-Verlag, Berlin 1982.

[45] W. Sierpiński, Sur l'equivalence de trois propriétés des ensembles abstraits, *Fund. Math.* 2 (1921), 179-188.

[46] W. Sierpiński, Sur un problème concernant les types de dimensions, *Fund. Math.* 19 (1932), 65-71.

[47] W. Sierpiński and A. Zygmund, Sur une fonction qui est discontinue sur tout ensemble de puissance du continu, *Fund. Math.* 4 (1923), 316-318.

[48] Z. Szentmiklóssy, S-spaces and L-spaces under Martin's axiom, *Coll. Math. Soc. János Bolyai* 23, Budapest 1978 (North-Holland, Amsterdam 1980), 1139-1145.

[49] Z. Szentmiklóssy, S-spaces can exist under MA, *Topology Appl.* 16 (1983), 243-251.

[50] F. Tall, The density topology, *Pacific J. Math.* 75 (1976), 275-284.

[51] M. G. Tkacenko, Chains and Cardinals, *Soviet Math. Doklady* 19 (1978), 382-385.

[52] S. Todorcevic, Cardinal Functions on Linearly Ordered Topological Spaces, in *Topology and Order Structures Part I*, (H.R. Bennet and D.J. Lutzer, eds.), Math. Centrum, Tract 142, Amsterdam (1981), 177-179.

[53] S. Todorcevic, On a Conjecture of R. Rado, *J. London Math. Soc.* (2) 27(1983), 1-8.

[54] S. Todorcevic, Forcing positive partition relations, *Trans. Amer. Math. Soc.* 280 (1983), 703-720.

[55] S. Todorcevic, A note on the Proper Forcing Axiom, *Contemporary Math.* 31 (1984), 209-218.

[56] S. Todorcevic, Trees and Linearly Ordered Sets, in *Handbook of Set-Theoretic Topology* (K. Kunen and J. Vaughan, eds.), Elsevier Science Publishers B.V. (1984), 235-293.

[57] S. Todorcevic, Directed Sets and Cofinal Types, *Trans. Amer. Math. Soc.* 290 (1985), 711-723.

[58] S. Todorcevic, Remarks on Chain Conditions in Products, *Compositio Math.* 55 (1985), 295-302.

[59] S. Todorcevic, Remarks on Cellularity in Products, *Compositio Math./* 57 (1986), 357-372.

[60] S. Todorcevic, Partitioning pairs of countable ordinals, *Acta Math.* 159 (1987), 261-294.

[61] S. Todorcevic, Oscillations of real numbers, *Logic Colloquium '86* (F.R. Drake and J.K. Truss, eds.), Elsevier Science Publishers (1988), 325-331.

[62] S. Todorcevic, Preprints:
    (1) (a) Generalized Luzin and Sierpiński sets, May 1983.
    (2) (b) Partition relations for uncountable cardinals less than or equal to the continuum, July 1983.
    (3) (c) hS and hL in finite powers, April 1984.
    (4) (d) Remarks on destructible partitions, July 1984.
    (5) (e) A combinatorial property of sets of irrationals, July 1984.
    (6) (f) A consequence of MA, December 1984.
    (7) (g) S $\neq$ L, July 1985.
    (8) (h) Luzin sets, February 1986.
    (9) (i) PFA and thin partitions, March 1986.
    (10) (j) Cellularity not attained, May 1986.
    (11) (k) Regular Radon measures are σ-finite, June 1986.
    (12) (ℓ) PFA and the continuum, August 1987.
    (13) (m) On a class of spaces associated with gaps, October 1987.
    (14) (n) A topology on sequences of countable ordinals, October 1988.
    (15) (o) Free sequences, October 1988.

[63] S. Todorcevic and B. Velickovic, Martin's axiom and partitions, *Compositio Math.* 63 (1987), 391-408.

[64] J. W. Tukey, Convergence and uniformity in topology, *Ann. of Math. Studies* 2 (Princeton Univ. Press, 1940).

[65] A. Weil, Sur les espaces à structure uniforme et sur la topologie géné-
     rale, *Actualités Sci. Indus.* 551, Paris 1937, Hermann et Cie.

[66] H. E. White, Topological spaces in which Blumberg's theorem holds,
     Proc. Amer. Math. Soc. 44 (1974), 454-462.

[67] P. Zenor, Hereditary m-separability and hereditary m-Lindelöf prop-
     erty in product spaces and function spaces, *Fund. Math.* 106 (1980),
     175-180.

[68] W. H. Woodin, Set theory and discontinuous homomorphisms from
     Banach algebras. Ph.D. thesis, University of California, Berkeley,
     1983.

[69] B. Shapirovskii, On discrete subspaces of topological spaces; weight,
     tightness and Souslin number, *Soviet Math. Dokl.* 13 (1972), 215-219.

[70] B. Shapirovskii, On separability and metrizability of spaces with
     Souslin condition, *Soviet Math. Dokl.* 13 (1972), 1633-1637.

[71] I. Juhász, Cardinal functions in topology–ten years later, *Math. Centre
     Tracts* 123, Amsterdam, 1980.

[72] K. Kunen, Strong S and L spaces under MA, in: *Set Theoretic Topology*
     (G.M. Reed, Ed.), Academic Press, 1977.

[73] K. Kunen and F. Tall, Between Martin's axiom and Souslin's Hypoth-
     esis, *Fund. Math.* 102 (1979), 173-181.

[74] A. Ya. Khinchin, *Continued fractions*, The University of Chicago
     Press, Chicago, 1964.

[75] A. Arhangel'skii, The construction and classification of topological
     spaces and cardinal invariants, *Russian Math. Surveys* 33 (1978), 33-
     96.

[76] A. Hajnal and I. Juhász, On hereditarily α-Lindelöf and hereditarily
     α-separable spaces, *Ann. Univ. Sci. Budapest* 11 (1968), 115-124.

[77] A. Hajnal and I. Juhász, Discrete subspaces of topological spaces II,
     Indag. Math. 31 (1969), 18-30.

[78] D. Kurepa, Sur les relations d'ordre, *Bull. Internat. Acad. Yougoslave*
     32 (1939), 66-76.

[79] S. Negrepontis, Banach Spaces and Topology, in *Handbook of Set Theoretic Topology* (K. Kunen and J. Vaughan, eds.), Elsevier Science Publishers B.V. (1984), 1045-1142.

[80] K. Kunen, A compact L-space, *Topology Appl* 12 (1981), 283-287.

[81] V. Malyhin, On tightness and Souslin number in exp X and in a product of spaces, *Soviet Math. Doklady* 13 (1972), 496-499.

[82] T. C. Przymusiński, Products of Normal Spaces, in *Handbook of Set Theoretic Topology* (K. Kunen and J. Vaughan, eds.), Elsevier Science Publishers B.V. (1984), 781-826.

[83] I Juhász, Cardinal Functions in Topology, *Math Centre Tracts* 34, Amsterdam (1971).

[84] W. W. Comfort and S. Negrepontis, *Chain Conditions in Topology*, Cambridge University Press, London (1982).

[85] K. Alster and P. Zenor, *An example concerning the preservation of the Lindelöf property in product spaces in Set-theoretic Topology* (G.M. Reed, ed.), Academic Press, New York (1977), 1-10.

[86] T. C. Przymusiński, Products of perfectly normal spaces, *Fund. Math.* 108 (1980), 129-136.

[87] A. W. Miller, Special Subsets of the Real Line, in *Handbook of Set Theoretic Topology* (K. Kunen and J. Vaughan, eds.), Elsevier Science Publishers B.V. (1984), 201-233.

[88] A. Hajnal and I. Juhász, Weakly separated subspaces and networks, *Logic Colloquium* 78 (M. Boffa et al., eds.), North Holland, Amsterdam (1979), 235-245.

[89] K. Ciesielski, Martin's axiom and a regular topological space with uncountable net weight whose countable product is hereditarily separable and hereditarily Lindelöf, *J. Symbolic Logic* 52 (1987), 396-399.

[90] A. V. Arhangel'skii, On cardinal invariants, in *General Topology and its Relations to Modern Analysis and Algebra III* (J. Novak, ed.), Proc. 3rd Prague Topological Symposium 1971, (Academia, Prague (1972)), 37-46.

[91] S. Todorcevic, A model with no S-spaces, preprint, March 1981.

[92] Z. Balogh, On compact Hausdorff spaces of countable tightness, preprint, 1987.

[93] D. Fremlin, Perfect preimages of ω_1 and the PFA, preprint, 1986.

[94] A. Dow, Removing large cardinals from the Moore-Mrowka problem, preprint, 1988.

[95] B. E. Shapirovskii, Spaces with Souslin and Shanin properties, *Mathematical Notes* 15 (1974), 161-164.

[96] B. E. Shapirovskii, Cardinal Invariants in Compact Hausdorff spaces, *Amer. Math. Soc. Transl.* (2) Vol. 134 (1987), 93-118.

[97] G. P. Amirdzanov and B. E. Shapirovskii, On everywhere dense subsets of topological spaces, *Soviet Math. Doklady* 15 (1974), 87-92.

[98] P. Erdös, F. Galvin and A. Hajnal, On set-systems having large chromatic number and not containing prescribed subsystems, *Coll. Math. Soc. Janos Bolyai, 10. Infinite and Finite Sets* (A. Hajnal, R. Rado and V. T. Soś, eds.) (North Holland, Amsterdam, 1975), 425-513.

[99] J. Steprans and S. Watson, Extending ideals, *Israel J. Math.* 54 (1986), 201-226.

[100] S. Todorcevic, Aronszajn trees and partitions, *Israel J. Math.* 52 (1985), 53-58.

[101] R. H. Sorgenfrey, On the topological product of paracompact spaces, *Bull. Amer. Math. Soc.* 53 (1947), 631-632.

[102] P. Mahlo, Über Teilmengen des Kontinuums von dessen Machtigkeit, Sitzungsberichte der Sachsischen Akad. der Wissenschaften zu Leipzig, *Math.-Naturwissenschaftliche* Klasse 65 (1913), 283-315.

[103] N. Luzin, Sur un problème de M. Baire, *C. R. Hebdomadaires Seances Acad. Sci. Paris*, 158 (1914), 1258-1261.

[104] H. E. White, Jr., Some Baire spaces for which Blumberg's theorem does not hold, *Proc. Amer. Math. Soc.* 51 (1975), 477-482.

[105] J. C. Oxtoby, *Measure and Category*, Springer-Verlag, New York, Heidelberg, Berlin, 1971.

[106] S. Mardesic and A. Prasolov, *Strong homology is not additive*, preprint 1986.

[107] A. Dow, P. simon and J. Vaughan, *Strong homology and the Proper Forcing Axiom*, preprint 1987.

[108] J. K. Truss, The noncomutativity of random and generic extensions, *J. Symbolic Logic* 48 (1983), 1009-1012.

[109] J. Chicón and J. Pawlikowksi, On ideals of subsets of the plane and on Cohen reals, *J. Symbolic Logic* 51 (1986), 560-569.

INDEX OF SYMBOLS

INDEX OF TERMS

ABCDEFGHIJ — 89